Phytogenic and Phytochemical as Alternative Feed Additives for Animal Production

Edited by

Youssef A. Attia
Sustainable Agriculture Production Research Group
Agriculture Department, Faculty of Environmental Sciences
King Abdulaziz University, Jeddah-21589
Saudi Arabia

Animal and Poultry Production Department
Faculty of Agriculture, Damanhour University
Damanhour-22713, Egypt

Mohamed E. Abd El-Hack
Poultry Department, Faculty of Agriculture
Zagazig University, Zagazig-44519
Egypt

Mahmoud M. Alagawany

Poultry Department, Faculty of Agriculture
Zagazig University, Zagazig-44511
Egypt

&

Asmaa Sh. Elnaggar

Animal and Poultry Production Department
Faculty of Agriculture, Damanhour University
Damanhour-22713, Egypt

Phytogenic and Phytochemical as Alternative Feed Additives for Animal Production

Editors: Youssef A. Attia, Mohamed E. Abd El-Hack, Mahmoud M. Alagawany & Asmaa Sh. Elnaggar

ISBN (Online): 978-981-5322-76-7

ISBN (Print): 978-981-5322-77-4

ISBN (Paperback): 978-981-5322-78-1

need for a court order if at any point you breach any terms of this License Agreement. In no event will any delay or failure by Bentham Science Publishers in enforcing your compliance with this License Agreement constitute a waiver of any of its rights.

3. You acknowledge that you have read this License Agreement, and agree to be bound by its terms and conditions. To the extent that any other terms and conditions presented on any website of Bentham Science Publishers conflict with, or are inconsistent with, the terms and conditions set out in this License Agreement, you acknowledge that the terms and conditions set out in this License Agreement shall prevail.

Bentham Science Publishers Pte. Ltd.
80 Robinson Road #02-00
Singapore 068898
Singapore
Email: subscriptions@benthamscience.net

BENTHAM SCIENCE

CONTENTS

Youssef A. Attia, Mohamed E. Abd El-Hack, Mahmoud M. Alagawany, Salem R. Alyileili, Khalid A. Asiry, Saber S. Hassan, Asmaa Sh. Elnaggar, Hany I. Habiba and *Shatha I. Alqurashi*

Youssef A. Attia, Mohamed E. Abd El-Hack, Mahmoud M. Alagawany, Mohamed A. AlBanoby and *Rashed A. Alhotan*

FOREWORD

In recent years, the field of animal nutrition has witnessed a transformative shift towards more sustainable and eco-friendly practices. The growing concern over antibiotic resistance, the demand for natural and safe food products, and the overarching need for environmental sustainability have driven researchers and practitioners to explore alternative feed additives. It is within this context that "Phytogenic and Phytochemical as Alternative Feed Additives for Animal Production", written by a fantastic group of authors, and many other highly respected colleagues, emerges as an essential contribution to the scientific literature and practical applications in animal nutrition.

Phytogenic substances derived from plants are a promising alternative to conventional feed additives. Their natural origins and diverse bioactive properties make them invaluable for promoting animal health, enhancing growth performance, and ensuring feed and food safety. This book provides a comprehensive exploration of various phytogenics and phytochemicals and details their roles, benefits, and practical applications in animal nutrition.

The first chapter, "Phytogenic Substances as Novel Feed Supplements and their Application in Livestock Nutrition," sets the stage by introducing the fundamental concepts and potential of phytogenics in revolutionizing animal feed. This chapter underscores the importance of integrating natural substances into animal diets to foster sustainable and healthy animal production systems.

Following this, the chapter on "Phytobiotics in Animal Nutrition" delves deeper into the specific categories and mechanisms by which these plant-derived compounds exert their beneficial effects. This discussion extends beyond mere supplementation, encompassing broader implications for animal health and productivity.

Subsequent chapters provide an in-depth examination of specific phytogenic substances, such as thyme, rosemary, milk thistle, turmeric, oregano essential oils, ginger, bee pollen, and propolis. Each chapter offers detailed insights into the unique properties, modes of action, and practical applications of these remarkable plant-based and natural additives. For instance, the therapeutic and antimicrobial properties of thyme and rosemary essential oils, the hepatoprotective effects of milk thistle, the anti-inflammatory and antioxidant benefits of turmeric, and the multifaceted health-promoting attributes of ginger, bee pollen, and propolis have been thoroughly explored.

As an academic and practitioner dedicated to advancing sustainable and health-oriented approaches in animal production, this book is a timely and invaluable resource. This brings together the latest research findings, practical insights, and real-world applications, providing a holistic understanding of how phytogenics and phytochemicals can be harnessed to achieve more sustainable and resilient animal production systems.

I was delighted when I received a request to write a brief foreword to the reprint of this book because, for several years, I have admired authors for incredible work. I commend the authors for their rigorous research and thoughtful presentation on this critical subject. This book will undoubtedly serve as a cornerstone reference for researchers, practitioners, students, and policymakers keen to foster a more sustainable future in animal nutrition.

With great enthusiasm, I invite you to delve into this comprehensive guide and explore the vast potential of phytogenics and phytochemicals to enhance animal health and production.

Nikola Puvača
Department of Engineering Management in Biotechnology
Faculty of Economics and Engineering Management in Novi Sad
University Business Academy in Novi Sad, Novi Sad, Serbia

PREFACE

Phytogenic feed additives (PFAs) have emerged as significant substitutes for antibiotics in animal nutrition, thereby addressing the urgent need for alternative growth promoters and health enhancers in livestock production. Plant-derived substances, including phytogenic substances and their bioactive components such as essential oils, flavonoids, and saponins, have been used to enrich growth rates, feed utilization, gut health, and overall animal immunity, productivity, and health. However, the ban on antibiotics as growth promoters owing to their contribution to antimicrobial resistance, carry-over effects, and health concerns has necessitated the search for alternatives; PFAs have gained attention for their multifunctional benefits. They not only promote growth, but also enhance gut health, modulate microbiota, and improve oxidative status without adverse effects associated with antibiotic use. The efficacy of specific PFAs such as thyme, curcumin, milk thistle, rosemary, bee pollen, and propolis has been highlighted, demonstrating potent antioxidant and antimicrobial activities. The roles and importance of PFAs in animal nutrition, product safety, and quality are multifaceted. They are viable alternatives to antibiotics and contribute to sustainable livestock production and consumer health. Continued research and development of PFAs are crucial for optimizing their use and realizing their full potential in various animal species. This book provides up-to-date reviews on valuable natural phytogenic growth promoters to completely or partially replace antibiotics as classical growth promoters, thus decreasing the adverse effects of antibiotics on animals, humans, and the environment. The book also supplies the scientific basis of phytogenic additives and services to scientists, students, the livestock industry, and the feed and food sectors for their efforts to improve animal and human health, product quality, and food safety.

Youssef A. Attia
Sustainable Agriculture Production Research Group
Agriculture Department, Faculty of Environmental Sciences
King Abdulaziz University, Jeddah-21589
Saudi Arabia

Animal and Poultry Production Department
Faculty of Agriculture, Damanhour University
Damanhour-22713, Egypt

Mohamed E. Abd El-Hack
Poultry Department, Faculty of Agriculture
Zagazig University, Zagazig-44519
Egypt

Mahmoud M. Alagawany
Poultry Department, Faculty of Agriculture
Zagazig University, Zagazig-44511
Egypt

&

Asmaa Sh. Elnaggar
Animal and Poultry Production Department
Faculty of Agriculture, Damanhour University
Damanhour-22713, Egypt

List of Contributors

Asmaa F. Khafaga
Department of Pathology, Faculty of Veterinary Medicine, Alexandria University, Apis, Alexandria, 21944, Egypt

Abdulmohsen H. Alqhtani
Department of Animal Production, College of Food and Agricultural Sciences, King Saud University, Riyadh, Saudi Arabia

Adel D. Al-qurashi
Sustainable Agriculture Production Research Group, Agriculture Department, Faculty of Environmental Sciences, King Abdulaziz University, Jeddah-21589, Saudi Arabia

Ayman E. Taha
Department of Animal Husbandry and Animal Wealth Development, Faculty of Veterinary Medicine, Alexandria University, Apis, Alexandria, 21944, Egypt

Ahmed A. Abdallah
Department of Poultry Nutrition, Animal Production Institute, Agricultural Research Center, Dokki, Giza-3751310, Egypt

Asmaa Sh. Elnaggar
Animal and Poultry Production Department, Faculty of Agriculture, Damanhour University, Damanhour-22713, Egypt

Bahaa Abou-Shehema
Department of Poultry Nutrition, Animal Production Institute, Agricultural Research Center, Dokki, Giza-3751310, Egypt

Fulvia Bovera
Department of Veterinary Medicine and Animal Production, University of Napoli Federico II, Via F. Delpino 1, 80137 Napoli, Italy

Hafez M. Hafez
Institute of Poultry Diseases, Faculty of Veterinary Medicine, Free University of Berlin, 14163, Berlin, Germany

Hossam A. Shahba
Rabbit, Turkey and Water Fowl Research Department, Animal Production Research Institute, Agriculture Research Center, Dokki, Giza-3751310, Egypt

Hany I. Habiba
Animal and Poultry Production Department, Faculty of Agriculture, Damanhour University, Damanhour-22713, Egypt

Khalid A. Asiry
Sustainable Agriculture Production Research Group, Agriculture Department, Faculty of Environmental Sciences, King Abdulaziz University, Jeddah-21589, Saudi Arabia

Mahmoud M. Alagawany
Poultry Department, Faculty of Agriculture, Zagazig University, Zagazig-44519, Egypt

Mohamed E. Abd El-Hack
Poultry Department, Faculty of Agriculture, Zagazig University, Zagazig-44519, Egypt

Mohamed A. AlBanoby
Al-Shamel Animal Feed Factory, Industrial Area, Hail-55411, Saudi Arabia

Maria Cristina de Oliveira
Faculty of Veterinary Medicine, University of Rio Verde, Rio Verde, GO 75.901-970, Brazil

Mayada R. Farag
Forensic Medicine and Toxicology Department, Veterinary Medicine Faculty, Zagazig University, Zagazig-44519, Egypt

Mohammed A. E. Naiel
Department of Animal Production, Faculty of Agriculture, Zagazig University, Zagazig-44511, Egypt

Mahmoud Madkour
Animal Production Department, National Research Centre, Giza, Egypt

Mohamed W. Radwan	Animal and Poultry Production Department, Faculty of Agriculture, Damanhour University, Damanhour-22713, Egypt
Nicola F. Addeo	Department of Veterinary Medicine and Animal Production, University of Napoli Federico II, Via F. Delpino 1, 80137 Napoli, Italy
Nehal K. Bertu	Directorate of Agriculture, Animal Production Department, Beheira Governorate, Damanhour, Egypt
Omer H.M. Ibrahim	Sustainable Agriculture Production Research Group, Agriculture Department, Faculty of Environmental Sciences, King Abdulaziz University, Jeddah-21589, Saudi Arabia Department of Floriculture, Faculty of Agriculture, Assiut University, Egypt
Rashed A. Alhotan	Department of Animal Production, College of Food and Agricultural Sciences, King Saud University, Riyadh, Saudi Arabia
Salem R. Alyileili	Department of Laboratory Analyses, College of Food and Agriculture Sciences, United Arab Emirates University, AlAin United Arab Emirates
Saber S. Hassan	Animal and Poultry Production Department, Faculty of Agriculture, Damanhour University, Damanhour-22713, Egypt
Shatha I. Alqurashi	Department of Biological Science, College of Science, University of Jeddah, Jeddah-21589, Saudi Arabia
Vincenzo Tufarelli	Department of Precision and Regenerative Medicine and Jonian Area (DiMePRe-J), Section of Veterinary Science and Animal Production, University of Bari Aldo Moro, 70010 Valenzano, Bari, Italy
Youssef A. Attia	Sustainable Agriculture Production Research Group, Agriculture Department, Faculty of Environmental Sciences, King Abdulaziz University, Jeddah-21589, Saudi Arabia Animal and Poultry Production Department, Faculty of Agriculture, Damanhour University, Damanhour-22713, Egypt

CHAPTER 1

Introduction and Background

Youssef A. Attia[1,3,*], **Mahmoud M. Alagawany**[2] and **Mohamed E. Abd El-Hack**[2]

[1] *Sustainable Agriculture Production Research Group, Agriculture Department, Faculty of Environmental Sciences, King Abdulaziz University, Jeddah-21589, Saudi Arabia*

[2] *Poultry Department, Faculty of Agriculture, Zagazig University, Zagazig-44519, Egypt*

[3] *Animal and Poultry Production Department, Faculty of Agriculture, Damanhour University, Damanhour-22713, Egypt*

Abstract: For centuries, plant-based ingredients extracted from herbs, spices, and medicinal flora have been used to enhance feed quality, flavor, and preservation, as well as in traditional healing medicine. As our understanding of their functional mechanisms grows, new opportunities arise for their application in the treatment of metabolic disorders and as feed supplements to promote positive physiological responses in various animal species. These naturally derived products are environmentally friendly and safe for living organisms, offering a wide range of beneficial properties, including antimicrobial, antioxidant, antiallergic, anticancer, antimutagenic, liver-protective, and immunomodulatory effects. Following the European Union's prohibition on antibiotic use as growth promoters in food-producing animals in 2006, researchers have turned their attention to natural alternatives, such as phytogenic substances, also referred to as phytobiotics or botanicals. These compounds have been demonstrated to boost animal productivity, encourage feed consumption, enhance nutrient absorption, and support optimal intestinal health. Promising feed additives include medicinal plants, such as milk thistle seeds, turmeric, rosemary leaves, and thyme. Additionally, bee pollen and propolis, which have both plant and animal origins, have been explored as substitutes for antibiotics and coccidiostats in animal nutrition, and have shown potential as growth enhancers and immune boosters. This book provides a comprehensive overview of the most commonly used natural substances as alternatives to growth-promoting antibiotics and details their mechanisms of action and effects in animals. The aim is to update the current knowledge and promote further research to identify additional beneficial natural molecules that can help reduce the negative impacts of antibiotics on animals, humans, and the environment.

Keywords: Animal nutrition, Antibiotic alternatives, Natural growth promoters (NGPs), Phytogenic feed additives.

* **Corresponding author Youssef A. Attia:** Sustainable Agriculture Production Research Group, Agriculture Department, Faculty of Environmental Sciences, King Abdulaziz University, Jeddah-21589, Saudi Arabia and Animal and Poultry Production Department, Faculty of Agriculture, Damanhour University, Damanhour-22713, Egypt; E-mail: yaattia@kau.edu.sa

BACKGROUND

Since their discovery, antibiotics have been tested and used for growth promotion when incorporated into the diet at sub-therapeutic levels. In addition, it is strictly related to the antimicrobial effect, which is normally used for disease control and/or prevention. The use of antibiotics as growth-promoting agents in food-producing animals was banned in the European Union in the beginning of 2006 because of cross-resistance and possible residues in animal products. This has prompted researchers to search for natural alternatives to antibiotics.

INTRODUCTION

Herbs, spices, and medicinal plants contain phytogenic ingredients that have been essential to human and animal health since ancient times. They have been used mainly to improve feed quality, taste, and preservation and in flak medicine for centuries. Growing awareness and understanding of the mechanisms of action of phytogenic constituents may provide opportunities to develop therapeutic interventions for metabolic diseases and their use as feed additives to stimulate positive physiological activities in mammals, birds, and fish. Phytogenic plants are a good source of diverse types of antioxidants and bioactive molecules. Additionally, natural products are safe for both living organisms and the environment. However, their compositions vary widely owing to their botanical origin, agronomical and environmental factors, and processing methods. Medicinal herbs and their derivatives possess numerous antimicrobial, antioxidative, antiallergic, anticancer, antimutagenic, hepatoprotective, and immunomodulatory properties.

Phytogenic substances, called phytobiotics or botanicals, are a group of natural growth promoters (NGPs) used as feed additives and are derived from herbs or other plants, but also products of animal origin or, better, with a double vegetable and animal origin. In the past, these products have been used to enhance animal productivity, stimulate feed intake, and improve nutrient digestibility, thus improving the feed conversion ratio in farms. These effects are believed to be due to increased endogenous digestive enzyme secretion, nutrient absorption, antioxidant immune stimulation, and antimicrobial and anthelmintic properties. However, when the various products were studied in greater depth, and therefore their mechanisms of action were known more precisely and in detail, it became clear that they can also have a positive effect on the animal's health, in particular exerting a positive effect on the intestinal microbiota. Today, it is well known that the intestinal environment represents an important barrier to the entry of pathogens into an organism; therefore, maintaining a good health status of the intestine contributes to maintaining good health and well-being.

Medicinal plants are used as growth enhancers in animal production to boost economic performance, improve the gut ecosystem, increase milk yield, enhance food quality, and improve fertility. Milk thistle seeds (*Silybum marianum L. Gaert., Asteraceae*), turmeric (curcumin), rosemary leaves (*Rosmarinus officinalis L.*) and thyme (*Thymus vulgaris*) are possible feed additives that may have promising effects.

Bee pollen (BP) and propolis (Pro) are products of double origin, and the raw materials (pollen or propolis) are of vegetable origin. However, both raw materials are partially modified by honeybees, which, once collected, mix them with their salivary secretions. Therefore, the positive effects of BP and Pro are due to both the raw materials and secretions of the honey bee. BP and Pro have been investigated as alternatives to antibiotics and coccidiostats in animal nutrition. Some studies have shown that these substances can be used as growth promoters and/or immune enhancers instead of antibiotics, as both have very interesting antioxidant activity.

This book provides an overview of the most common natural substances currently used as alternatives to growth-promoting antibiotics. The description of the different substances, their mechanisms of action, and their effects on animals aims to provide a precise and updated picture of the state of the art, with the aim of stimulating further research in the identification of other natural molecules with benefits that could be present in the plant world and beyond, to decrease the adverse effects of antibiotics on animals, humans, and the environment.

CHAPTER 2

Phytogenic Substances as Novel Feed Supplements and their Application in Livestock Nutrition

Youssef A. Attia[1,2,*], **Nicola F. Addeo**[3], **Fulvia Bovera**[3], **Mohamed E. Abd Al-Hack**[4], **Mohamed A. AlBanoby**[5], **Rashed A. Alhotan**[6], **Asmaa F. Khafaga**[7], **Hafez M. Hafez**[8] and **Maria Cristina de Oliveira**[9]

[1] *Sustainable Agriculture Production Research Group, Agriculture Department, Faculty of Environmental Sciences, King Abdulaziz University, Jeddah-21589, Saudi Arabia*

[2] *Animal and Poultry Production Department, Faculty of Agriculture, Damanhour University, Damanhour-22713, Egypt*

[3] *Department of Veterinary Medicine and Animal Production, University of Napoli Federico II, Via F. Delpino 1, 80137 Napoli, Italy*

[4] *Poultry Department, Faculty of Agriculture, Zagazig University, Zagazig-44519, Egypt*

[5] *Al-Shamel Animal Feed Factory, Industrial Area, Hail-55411, Saudi Arabia*

[6] *Department of Animal Production, College of Food and Agricultural Sciences, King Saud University, Riyadh, Saudi Arabia*

[7] *Department of Pathology, Faculty of Veterinary Medicine, Alexandria University, Apis, Alexandria, 21944, Egypt*

[8] *Institute of Poultry Diseases, Faculty of Veterinary Medicine, Free University of Berlin, 14163, Berlin, Germany*

[9] *Faculty of Veterinary Medicine, University of Rio Verde, Rio Verde, GO 75.901-970, Brazil*

Abstract: Phytogenic substances derived from plant organs are bioactive compounds widely used to enhance health, food safety and shelf-life. These substances exhibit diverse biological activities, supporting animal and human health by promoting antioxidant defenses and enhancing immune function. They can mitigate the production of oxygen-containing reactive species (ROS) generated by environmental stressors, inhibit enzymes implicated in cellular damage, enhance mitochondrial function, and improve energy biosynthesis and availability. Incorporating phytogenic additives into livestock diets is a safe, effective, and economically viable strategy for mitigating the adverse effects of conventional feed additives on animal and human health. Reducing antibiotic use in livestock production is critical and can be achieved by integrating phytogenic substances, herbs, spices, medicinal plants, probiotics, prebiotics, synbio-

* **Corresponding author Youssef A. Attia:** Sustainable Agriculture Production Research Group, Agriculture Department, Faculty of Environmental Sciences, King Abdulaziz University, Jeddah-21589, Saudi Arabia & Animal and Poultry Production Department, Faculty of Agriculture, Damanhour University, Damanhour-22516, Egypt; E-mail: yaattia@kau.edu.sa

tics, and postbiotics along with implementing robust biosecurity measures. This chapter emphasizes the role of phytogenic products as growth promoters in livestock production, and their potential applications in enhancing food safety and security. The subtherapeutic use of antibacterial drugs has significantly enhanced meat, egg, and milk production over the centuries. However, the use of antimicrobial agents promotes the selection of resistant microbes that can proliferate rapidly and become dominant within microbial populations, potentially compromising the effectiveness of treatment for microbial infections in humans.

Keywords: Antioxidants, Bioactive substances, Immunity, Production enhancers, Phytogenic.

INTRODUCTION

Phytogenic plants are rich sources of antioxidants and bioactive compounds, making them suitable alternatives to antimicrobial agents because of their beneficial effects on animal nutrition [1]. Phytogenic substances are environmentally safe and compatible with sustainable livestock production. However, their efficacy can vary significantly depending on factors such as processing methods, botanical origin, and agronomic and environmental conditions [1, 2]. Medicinal plants, herbs, spices, and their derivatives contain a wide array of bioactive compounds with diverse properties including antioxidant, antimicrobial [3], anti-allergic [4], anticancer [5], anti-mutagenic [6], hepatoprotective [7], and immunomodulatory [8] effects. Fig. (**1**) presents a historical overview of the inhibition of antibiotics as growth promoters, which has driven global research efforts to identify natural alternatives. Among these, phytogenic products have emerged as prominent candidates for replacing antibiotics in livestock production.

Phytogenic substances, also referred to as phytobiotics or botanicals, are natural growth promoters (NGPs) that are used as feed supplements derived from herbs or other plants to enhance animal productivity and improve digestibility [9]. Their effects are believed to stem from increased endogenous digestive enzyme secretion, improved nutrient absorption, enhanced antioxidant activity, immune stimulation, and potential antimicrobial and anthelminthic properties [10, 11]. Medicinal plants have been utilized to boost livestock performance by improving gut microbiota, milk yield, food quality, and fertility [12 - 18]. Various plant-derived feed additives, such as thyme [19, 20], milk thistle seeds [21 - 24], rosemary leaves [25, 26], and turmeric, have demonstrated promising effects on livestock immunity and overall health, based on cost-benefit analyses [16]. Additionally, probiotics and prebiotics, such as fructooligosaccharides and mannan oligosaccharides (MOS), have emerged as effective non-antibiotic growth promoters [27 - 33]. Organic acids and their salts, as well as essential oils, are recognized as safe by the EU, and have been widely adopted as natural feed additives in animal production [34 - 41]. Bee pollen and propolis have also been explored as antibiotic and coccidiostat alternatives in animal nutrition, with studies indicating that both can serve as growth promoters and immune enhancers [42 - 49]. This chapter focuses on the current alternatives to antibiotic growth promoters (AGPs) and their potential applications in animal nutrition. The aim was to encourage further research on natural growth promoters to replace AGPs, thereby reducing the detrimental effects of antibiotics on animals, humans, and the environment.

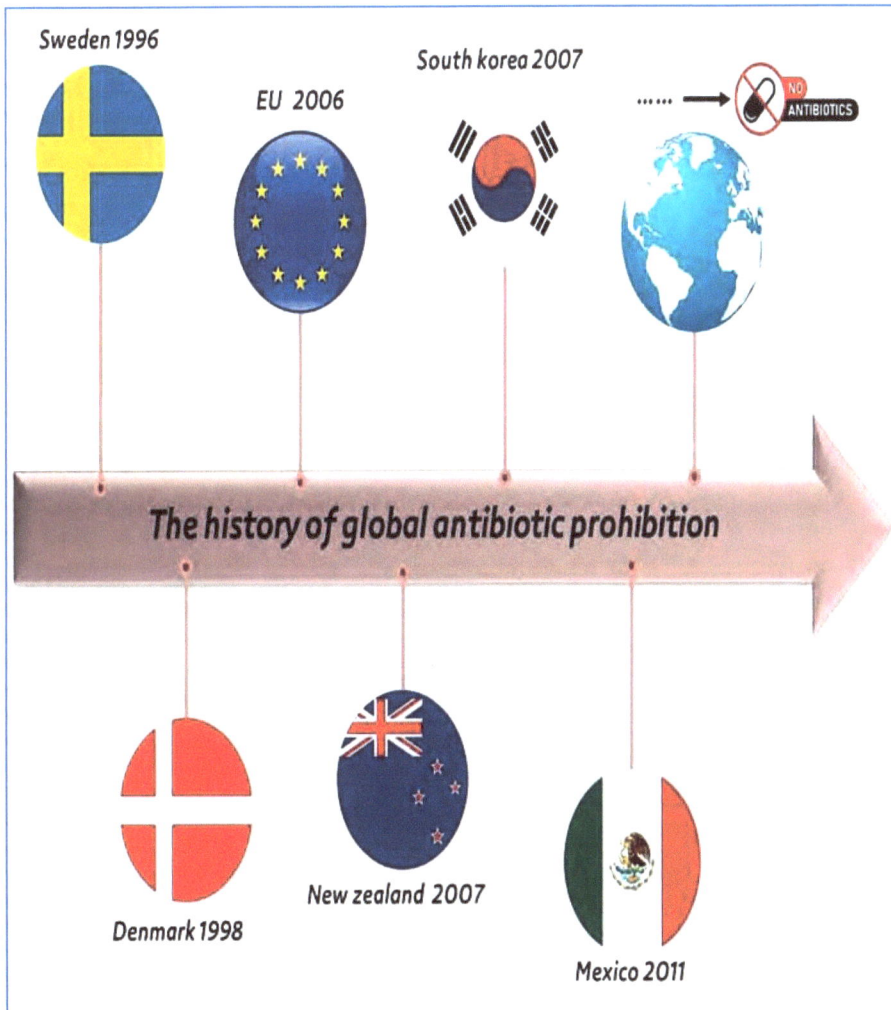

Fig. (1). The history of antibiotic prohibition.

ANTIBIOTICS

Stimulation of Growth Through the Use of Antimicrobial Substances

For several centuries, the subtherapeutic use of antibacterial drugs has significantly enhanced the production of meat, eggs, and milk [1, 50]. Among livestock, poultry has become one of the fastest-growing sources of animal protein worldwide, accounting for nearly a quarter of the total meat production over the past century. Modern poultry production systems are now capable of raising broiler chickens to market weight for as little as 30 d. This rapid progress has been driven by advancements in genetic selection, improved husbandry practices, optimized nutrition, and enhanced health management strategies.

Risks of Antibiotics as Growth Promoters

Antimicrobial agents are used in animal farming to treat infections and alleviate the negative effects of diseases on both animal and human health. Although the economic and health benefits of antibiotic use have driven a transformation in livestock production, these practices have also resulted in unintended consequences, including antibiotic residues in animal products, cross-resistance, and carryover effects [16]. Unfortunately, the use of antimicrobial agents promotes the selection of resistant microbes that proliferate rapidly and become dominant within microbial populations. These resistant microorganisms can pass on their advantageous traits to future generations and other microbial species through genetic mutations or plasmid-mediated gene transfer [51]. According to the World Health Organization (WHO), antimicrobial resistance (AMR) refers to the ability of microorganisms to withstand the effects of antimicrobial agents [52]. AMR can potentially be transmitted from livestock to humans through the consumption or handling of animal products contaminated with resistant microbes [53, 54]. Once acquired, these resistant microbes may establish themselves in the human gut, where the genes responsible for antibiotic resistance can spread to other bacteria within the human microbiome, compromising the effectiveness of treatments for microbial infections [55].

Development of Antibiotic Resistance

Microbes display remarkable adaptability, primarily because of their short reproductive cycles (as brief as 15-20 minutes under optimal conditions) and their ability to exchange genetic information across different species. While antimicrobial drugs can effectively eliminate most bacteria within a given environment, resistant strains can survive and proliferate, passing on their resistance traits to subsequent generations and potentially to other bacterial species. The widespread use of antibiotics in both human medicine and animal husbandry has accelerated the emergence of antibiotic-resistant strains, which pose significant risks to public health and livestock management. Although these resistant microbes do not directly affect human health, they are still concerning because of their ability to transfer resistant genes to pathogenic bacteria. Strains of bacteria resistant to antimicrobial drugs, including *Salmonella* spp., *Escherichia coli*, and *Campylobacter* spp., have been detected in livestock populations across numerous countries [57]. For example, the emergence of fluoroquinolone-resistant *Campylobacter jejuni* in chickens followed the introduction of fluoroquinolones into their feed [58]. The use of antimicrobial drugs in feed for food-producing animals can lead to several harmful consequences, such as interspecies and cross-species resistance to antimicrobials; residues of antibiotics in animal products such as eggs, meat, and milk; and disruptions to the natural bacterial balance [59 - 61]. Fig. (**2**) provides a global overview of antibiotic use in food-producing animals.

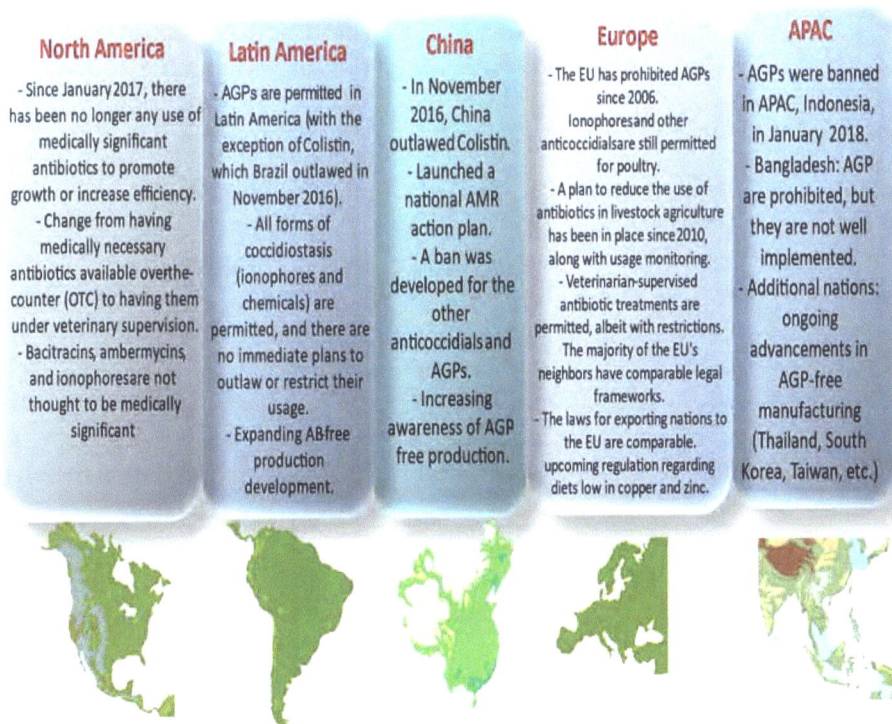

Fig. (2). A global overview of antibiotics used in livestock feed.

PHYTOGENIC FEED ADDITIVES

Characteristics of Botanical Feed Supplements

Phytogenic feed additives (PFAs) are plant-derived compounds recognized for their beneficial effects on animal health and performance. They are incorporated into animal diets as natural supplements to enhance overall health and production efficiency [62]. PFAs encompass a wide range of plant compounds with varying properties, sourced from herbs, spices (such as garlic, anise, cinnamon, thyme, and rosemary), essential oils, oleoresins [63], and fruit-based compounds like flavonoids [64]. The concentrations of active compounds in PFAs can vary considerably depending on factors such as the plant part used, harvest season, and geographical origin [62]. For instance, essential oils and volatile compounds, which are known for their antibacterial, antiviral, and antioxidant properties, are typically found in the seeds and leaves. However, their concentrations can be influenced by plant variety, seed or leaf maturity, and climatic conditions during growth [65]. These variations may affect the efficacy of PFAs in animal diets, necessitating careful consideration of their formulation and applications.

Effects of Dietary Supplementation with PFAs

It is frequently suggested that PFA supplementation can significantly influence the palatability and gut function of animals. PFAs can enhance feed palatability by imparting pleasant flavors and encouraging greater feed intake. However, few studies have specifically assessed the effects of PFAs on palatability using selective feeding patterns [62]. Maass *et al.* [66] and Roth-Maier *et al.* [67] reported that the inclusion of phytogenic compounds in broiler and layer hen diets resulted in reduced feed intake. Typically, increased feed consumption in poultry is associated with supplements such as prebiotics, probiotics, and organic acids [68]. Various PFAs, including spasmolytics and laxatives, are known to promote digestive health [69]. Moreover, PFAs have been shown to boost enzyme activity in the digestive tract and increase the secretion of bile, mucus, and saliva, potentially improving overall digestive efficiency [70]. Research has also indicated that PFAs may stimulate activity in the chicken intestinal mucosa, potentially affecting pathogen attachment and promoting a stable microbial balance in the gut [71]. Shen *et al.* [72] demonstrated that supplementation with 0.05% oregano essential oil significantly increased body weight and feed intake during the fattening stage and improved feed conversion during the starter phase in geese. Although the same concentration of oregano oil did not significantly alter the cecal microbiota at 87 days of age, it did influence the microbial populations at the genus level.

Antioxidant Action of Phytogenic Feed Additives

Herbs and spices are widely recognized for their antioxidant properties. Among various plants, volatile oils from the *Lamiaceae* family, particularly rosemary, have attracted considerable attention as natural antioxidants in both human and animal nutrition because of their phenolic terpenes, such as rosmarinic acid and carnosol. Other plants in this family, including thyme and oregano, are also rich in monoterpenes such as thymol and carvacrol, which possess antioxidant properties [73]. In broiler diets, thymol has been shown to reduce fatty acid oxidation, as evidenced by decreased malondialdehyde levels in the duodenal mucosa [74]. Additional plant species from the *Zingiberaceae* family (such as ginger and *turmeric*), *Umbelliferae* family (including anise and coriander), and green tea, which are rich in flavonoids, are also known for their antioxidant activities [75, 76]. However, whether phytogenic antioxidants can effectively replace conventional synthetic antioxidants commonly used in animal feed remains uncertain and warrants further investigation [62]. Recently, Oni *et al.* [77] reviewed the roles of bioactive compounds in PFAs, such as polyphenols, flavonoids, antioxidants, growth-promoting agents, and immune-modulating compounds, in the management of heat stress in broiler chickens.

Antimicrobial Actions of Phytogenic Feed Additives

Although herbs and spices are widely documented to exert significant antimicrobial effects against pathogens, their efficacy is believed to be closely related to the specific physicochemical properties of their active plant compounds [78]. Key constituents, particularly phenolic compounds, are primarily responsible for these antimicrobial activities, as previously noted for their antioxidant effects [79]. The antimicrobial efficacy of these compounds is thought to be largely due to their ability of hydrophobic essential oils to penetrate bacterial membranes, disrupt integrity, and cause ion leakage. Additionally, strong antibacterial effects have been observed for non-phenolic compounds such as limonene and isoquinoline alkaloids found in *Sanguinaria canadensis* [79, 80]. Studies on broilers have demonstrated the effectiveness of essential oils in combating pathogens like *Escherichia coli* and *Clostridium perfringens* [81, 82]. Literature suggests that in broilers, the antimicrobial efficacy of phytogenic compounds within living systems should not be overlooked. In addition to their antimicrobial properties, phytogenics also appear to improve carcass hygiene and preservation [83 - 85]. The European Food Safety Authority (EFSA) has recommended the incorporation of phytogenic alternatives as a potentially effective strategy to reduce bacterial contamination in animal nutrition and control the spread of foodborne pathogens. Plant extracts applied to food or carcass surfaces have been shown to reduce microbial contamination of poultry products [86]. Essential oils derived from plants, such as oregano, rosemary, and sage, have been reported to positively affect carcass hygiene [87, 88]. Jambwa *et al.* [89] observed that *Securidaca longepedunculata* (a plant used in poultry ethnomedicine in Zimbabwe) exhibited a range of therapeutic properties, including antibacterial activity against poultry pathogens, and can be considered a viable feed supplement in poultry diets. However, it is premature to fully assess the long-term effectiveness and reliability of these decontamination methods.

CONCLUSION

In conclusion, phytogenic substances have emerged as promising alternatives to conventional antibiotics in livestock production, offering a range of benefits for animal health, productivity, and food safety. These plant-derived compounds demonstrate diverse biological activities, including antioxidant, antimicrobial, and immune-enhancing properties, and can effectively support animal growth and welfare without risks associated with antibiotic use. The incorporation of phytogenic additives into animal diets represents a sustainable and economically viable approach to mitigate the adverse effects of conventional feed additives on both animal and human health. As the livestock industry continues to face challenges related to antibiotic resistance and consumer demand for antibiotic-free animal products, the integration of phytogenic substances with other natural alternatives such as probiotics and prebiotics offers an eco-friendly approach. However, it is important to note that the efficacy of these compounds can vary depending on factors, such as plant origin, processing methods, and environmental conditions. Therefore, further research is required to optimize the use of phytogenic additives, such as in conversion and organic animal farming, in different livestock species and production systems. The continued exploration and development of phytogenic feed additives will play a crucial role in promoting sustainable livestock production, enhancing food safety, and addressing global concerns regarding antimicrobial resistance. By harnessing the power of these natural compounds, the livestock industry can work towards a future that balances productivity with animal welfare and public health considerations.

IMPLICATIONS

- Feed additives can enhance gut health, boost immune function, and improve overall animal performance, without the risks associated with antibiotics.
- Decreasing reliance on antibiotics in livestock and the development and spread of antibiotic-resistant bacteria can be mitigated, thus benefiting both animal and human health.

- The application of phytogenic substances can improve carcass hygiene and reduce bacterial contamination in animal products, potentially reducing foodborne illnesses.
- Phytogenic additives offer an environmentally friendly and economically viable alternative to conventional growth promoters, aligning with the consumer demand for more natural production methods.
- A wide range of plant-derived compounds provides opportunities for tailored solutions to address specific health and productivity challenges in different livestock species and production systems.
- The variability in the efficacy of phytogenic substances highlights the need for further research to optimize their use and understand their long-term effects on animal health and production.
- The growing interest in phytogenic substances may drive innovation in feed formulation, processing techniques, and the development of new plant-based additives for the livestock industry.

REFERENCES

[1]   Castanon JIR. History of the use of antibiotic as growth promoters in European poultry feeds. Poult Sci 2007; 86(11): 2466-71.
[http://dx.doi.org/10.3382/ps.2007-00249] [PMID: 17954599]

[2]   Windisch W, Kroismayr A. The effects of phytobiotics on performance and gut function in monogastrics. World Nutrition Forum 2016; 85-90.

[3]   Parham S, Kharazi AZ, Bakhsheshi-Rad HR, *et al.* Antioxidant, antimicrobial and antiviral properties of herbal materials. Antioxidants 2020; 9(12): 1309.
[http://dx.doi.org/10.3390/antiox9121309] [PMID: 33371338]

[4]   Saini R, Dhiman NK. Natural anti-inflammatory and anti-allergy agents: herbs and botanical ingredients. Medicinal Chemistry - Anti-inflammatory & Anti-Allergy Agents 2022; 21: 90-114.
[http://dx.doi.org/10.2174/1871523021666220411111743]

[5]   Chandra S, Gahlot M, Choudhary AN, *et al.* Scientific evidences of anticancer potential of medicinal plants. Food Chem Adv 2023; 2: 100239.
[http://dx.doi.org/10.1016/j.focha.2023.100239]

[6]   Akram M, Riaz M, Wadood AWC, *et al.* Medicinal plants with anti-mutagenic potential. Biotechnol Biotechnol Equip 2020; 34(1): 309-18.
[http://dx.doi.org/10.1080/13102818.2020.1749527]

[7]   Ilyas U, Katare DP, Aeri V, Naseef PP. A review on hepatoprotective and immunomodulatory herbal plants. Pharmacognosy Reviews 2016; 10: 66-70. 4.
[http://dx.doi.org/10.4103/0973-7847.176544]

[8]   Alanazi HH, Elasbali AM, Alanazi MK, El Azab EF. Medicinal herbs: promising immunomodulators for the treatment of infectious diseases. Molecules 2023; 28(24): 8045.
[http://dx.doi.org/10.3390/molecules28248045] [PMID: 38138535]

[9]   Ayalew H, Zhang H, Wang J. Parham S, Kharazi AZ, Bakhsheshi-Rad HR, Nur H, Ismail AF, Sharif S, RamaKrishna S, Berto F. Potential feed additives as antibiotic alternatives in broiler production. Front Vet Sci 2022; 9: 916473.
[http://dx.doi.org/10.3389/fvets.2022.916473] [PMID: 35782570]

[10]   Upadhaya SD, Kim SJ, Kim IH. Effects of gel-based phytogenic feed supplement on growth

performance, nutrient digestibility, blood characteristics and intestinal morphology in weanling pigs. J Appl Anim Res 2016; 44(1): 384-9.
[http://dx.doi.org/10.1080/09712119.2015.1091334]

[11] Abdelli N, Solà-Oriol D, Pérez JF. Phytogenic feed additives in poultry: achievements, perspective and challenges. Animals (Basel) 2021; 11(12): 3471.
[http://dx.doi.org/10.3390/ani11123471] [PMID: 34944248]

[12] Orlowski S, Flees J, Greene ES, *et al.* Effects of phytogenic additives on meat quality traits in broiler chickens1. J Anim Sci 2018; 96(9): 3757-67.
[http://dx.doi.org/10.1093/jas/sky238] [PMID: 30184154]

[13] Kholif AE, Hassan AA, El Ashry GM, *et al.* Phytogenic feed additives mixture enhances the lactational performance, feed utilization and ruminal fermentation of Friesian cows. Anim Biotechnol 2021; 32(6): 708-18.
[http://dx.doi.org/10.1080/10495398.2020.1746322] [PMID: 32248772]

[14] Rashid Z, Mirani ZA, Zehra S, *et al.* Enhanced modulation of gut microbial dynamics affecting body weight in birds triggered by natural growth promoters administered in conventional feed. Saudi J Biol Sci 2020; 27(10): 2747-55.
[http://dx.doi.org/10.1016/j.sjbs.2020.06.027] [PMID: 32994734]

[15] Swelum AA, Hashem NM, Abdelnour SA, *et al.* Effects of phytogenic feed additives on the reproductive performance of animals. Saudi J Biol Sci 2021; 28(10): 5816-22.
[http://dx.doi.org/10.1016/j.sjbs.2021.06.045] [PMID: 34588896]

[16] Shehata AA, Yalçın S, Latorre JD, *et al.* Probiotics, prebiotics, and phytogenic substances for optimizing gut health in poultry. Microorganisms 2022; 10(2): 395.
[http://dx.doi.org/10.3390/microorganisms10020395] [PMID: 35208851]

[17] Yu SJ, Bajagai YS, Petranyi F, Stanley D. Phytogen improves performance during spotty liver disease by impeding bacterial metabolism and pathogenicity. Appl Environ Microbiol 2022; 88(18): e00758-22.
[http://dx.doi.org/10.1128/aem.00758-22] [PMID: 36094201]

[18] Mahasneh ZMH, Abuajamieh M, Abedal-Majed MA, Al-Qaisi M, Abdelqader A, Al-Fataftah ARA. Effects of medical plants on alleviating the effects of heat stress on chickens. Poult Sci 2024; 103(3): 103391.
[http://dx.doi.org/10.1016/j.psj.2023.103391] [PMID: 38242055]

[19] Ezzat Ahmed A, Alkahtani MA, Abdel-Wareth AAA. Thyme leaves as an eco-friendly feed additive improves both the productive and reproductive performance of rabbits under hot climatic conditions. Vet Med (Praha) 2020; 65(12): 553-63.
[http://dx.doi.org/10.17221/42/2020-VETMED]

[20] Ribeiro ADB, Ferraz Junior MVC, Polizel DM, *et al.* Effect of thyme essential oil on rumen parameters, nutrient digestibility, and nitrogen balance in wethers fed high concentrate diets. Arq Bras Med Vet Zootec 2020; 72(2): 573-80.
[http://dx.doi.org/10.1590/1678-4162-11322]

[21] Tedesco DEA, Guerrini A. Use of milk thistle in farm and companion animals: a review. Planta Med 2023; 89(6): 584-607.
[http://dx.doi.org/10.1055/a-1969-2440] [PMID: 36302565]

[22] Abdel-Latif HMR, Shukry M, Noreldin AE, *et al.* Milk thistle (*Silybum marianum*) extract improves growth, immunity, serum biochemical indices, antioxidant state, hepatic histoarchitecture, and intestinal histomorphometry of striped catfish, Pangasianodon hypophthalmus. Aquaculture 2023; 562: 738761.
[http://dx.doi.org/10.1016/j.aquaculture.2022.738761]

[23] Bencze-Nagy J, Strifler P, Horváth B, *et al.* Effects of dietary milk thistle (*Silybum marianum*) supplementation in ducks fed mycotoxin-contaminated diets. Vet Sci 2023; 10(2): 100.

[http://dx.doi.org/10.3390/vetsci10020100] [PMID: 36851404]

[24]   Elnesr SS, Elwan HAM, El Sabry MI, Shehata AM. The nutritional importance of milk thistle (*Silybum marianum*) and its beneficial influence on poultry. Worlds Poult Sci J 2023; 79(4): 751-68.
[http://dx.doi.org/10.1080/00439339.2023.2234339]

[25]   Khayyal A, El-Badawy M, Ashmawy T. Effect of rosemary or laurel leaves as feed additives on performance of growing lambs. Egypt J Nutr Feeds 2021; 24(3): 343-56.
[http://dx.doi.org/10.21608/ejnf.2021.210836]

[26]   Yao Y, Liu Y, Li C, *et al.* Effects of rosemary extract supplementation in feed on growth performance, meat quality, serum biochemistry, antioxidant capacity, and immune function of meat ducks. Poult Sci 2023; 102(2): 102357.
[http://dx.doi.org/10.1016/j.psj.2022.102357] [PMID: 36502565]

[27]   Markowiak P, Śliżewska K. The role of probiotics, prebiotics and synbiotics in animal nutrition. Gut Pathog 2018; 10(1): 21.
[http://dx.doi.org/10.1186/s13099-018-0250-0] [PMID: 29930711]

[28]   Shang Y, Kumar S, Thippareddi H, Kim WK. Effect of dietary fructooligosaccharide (FOS) supplementation on ileal microbiota in broiler chickens. Poult Sci 2018; 97(10): 3622-34.
[http://dx.doi.org/10.3382/ps/pey131] [PMID: 30016495]

[29]   Hazrati S, Rezaeipour V, Asadzadeh S. Effects of phytogenic feed additives, probiotic and mannan-oligosaccharides on performance, blood metabolites, meat quality, intestinal morphology, and microbial population of Japanese quail. Br Poult Sci 2020; 61(2): 132-9.
[http://dx.doi.org/10.1080/00071668.2019.1686122] [PMID: 31661976]

[30]   Abd El-Hack ME, El-Saadony MT, Shafi ME, *et al.* Probiotics in poultry feed: A comprehensive review. J Anim Physiol Anim Nutr (Berl) 2020; 104(6): 1835-50.
[http://dx.doi.org/10.1111/jpn.13454] [PMID: 32996177]

[31]   Arsène MMJ, Davares AKL, Andreevna SL, *et al.* The use of probiotics in animal feeding for safe production and as potential alternatives to antibiotics. Vet World 2021; 14(2): 319-28.
[http://dx.doi.org/10.14202/vetworld.2021.319-328] [PMID: 33776297]

[32]   Chang L, Ding Y, Wang Y, *et al.* Effects of different oligosaccharides on growth performance and intestinal function in broilers. Front Vet Sci 2022; 9: 852545.
[http://dx.doi.org/10.3389/fvets.2022.852545] [PMID: 35433897]

[33]   Obianwuna UE, Chang XY, Wang J, *et al.* Dietary fructooligosaccharides effectively facilitate the production of high-quality eggs *via* improving the physiological status of laying hens. Foods 2022; 11(13): 1828.
[http://dx.doi.org/10.3390/foods11131828] [PMID: 35804644]

[34]   Azza MK, Naela M. Ragaa. Effect of dietary supplementation of organic acids on performance and serum biochemistry of broiler chicken. Nat Sci 2024; 12: 38-45.
[https://www.sciencepub.net/nature/ns1202/00622976ns120214_38_45.pdf].

[35]   Zeng Z, Zhang S, Wang H, Piao X. Essential oil and aromatic plants as feed additives in non-ruminant nutrition: a review. J Anim Sci Biotechnol 2015; 6(1): 7.
[http://dx.doi.org/10.1186/s40104-015-0004-5] [PMID: 25774291]

[36]   Stevanović ZD, Bošnjak-Neumüller J, Pajić-Lijaković I, Raj J, Vasiljević M. Essential oils as feed additives – future perspectives. Molecules 2018; 23(7): 1717.
[http://dx.doi.org/10.3390/molecules23071717] [PMID: 30011894]

[37]   Nguyen DH, Kim IH. Protected organic acids improved growth performance, nutrient digestibility, and decreased gas emission in broilers. Animals (Basel) 2020; 10(3): 416.
[http://dx.doi.org/10.3390/ani10030416] [PMID: 32131472]

[38]   Kholif AE, Olafadehan OA. Essential oils and phytogenic feed additives in ruminant diet: chemistry, ruminal microbiota and fermentation, feed utilization and productive performance. Phytochem Rev

2021; 20(6): 1087-108.
[http://dx.doi.org/10.1007/s11101-021-09739-3]

[39]   Ferreira TS, Ravetti R, Rubio MS, *et al.* Inclusion of organic acids in the drinking water and feed for the control of *Salmonella Heidelberg* in broilers. Brazilian Poultry Science Journal 2022; 24: eRBCA-2020-1427.
[http://dx.doi.org/10.1590/1806-9061-2020-1427]

[40]   Khan RU, Naz S, Raziq F, *et al.* Prospects of organic acids as safe alternative to antibiotics in broiler chickens diet. Environ Sci Pollut Res Int 2022; 29(22): 32594-604.
[http://dx.doi.org/10.1007/s11356-022-19241-8] [PMID: 35195862]

[41]   Qui NH. Recent advances of using organic acids and essential oils as in-feed antibiotic alternative in poultry feeds. Czech J Anim Sci 2023; 68(4): 141-60.
[http://dx.doi.org/10.17221/99/2022-CJAS]

[42]   Dias DMB, Oliveira MC, Silva DM, Bonifácio NP, Claro DC, Marchesin WA. Bee pollen supplementation in diets for rabbit does and growing rabbits. Animal Production –. Anim Sci 2023; 35: 425-30.
[http://dx.doi.org/10.4025/actascianimsci.v35i4.18950]

[43]   Abdelnour SA, Abd El-Hack ME, Alagawany M, Farag MR, Elnesr SS. Beneficial impacts of bee pollen in animal production, reproduction and health. J Anim Physiol Anim Nutr (Berl) 2019; 103(2): 477-84.
[http://dx.doi.org/10.1111/jpn.13049] [PMID: 30593700]

[44]   Oliveira MC, Souza RG, Dias DMB, Gonçalves BN. Bee pollen improves productivity of laying Japanese quails. Rev Bras Saúde Prod Anim 2020; 21: e212135020.
[http://dx.doi.org/10.1590/s1519-99402121352020]

[45]   Pieroni CA, Oliveira MC, Santos WLR, Mascarenhas LB, Oliveira MAD. Effect of green propolis on the productivity, nutrient utilisation, and intestinal morphology of Japanese laying quail. Rev Bras Zootec 2020; 49: e20190198.
[http://dx.doi.org/10.37496/rbz4920190198]

[46]   Morsy AS, Soltan YA, El-Zaiat HM, Alencar SM, Abdalla AL. Bee propolis extract as a phytogenic feed additive to enhance diet digestibility, rumen microbial biosynthesis, mitigating methane formation and health status of late pregnant ewes. Anim Feed Sci Technol 2021; 273: 114834.
[http://dx.doi.org/10.1016/j.anifeedsci.2021.114834]

[47]   Nemauluma MFD, Manyelo TG, Ng'ambi JW, Malematja EM, Kolobe SD. Effects of bee pollen inclusion on the performance and gut morphology of Ross 308 broiler chickens. Brazilian Journal of Poultry Science 2023; 25: eRBCA-2022-1632.
[http://dx.doi.org/10.1590/1806-9061-2022-1632]

[48]   Santos EL, Barbosa JM, Porto-Neto FF, *et al.* Propolis extract as a feed additive of the Nile tilapia juveniles. Arq Bras Med Vet Zootec 2023; 75(4): 744-52.
[http://dx.doi.org/10.1590/1678-4162-12806]

[49]   Sierra-Galicia MI, Rodríguez-de Lara R, Orzuna-Orzuna JF, Lara-Bueno A, Ramírez-Valverde R, Fallas-López M. Effects of supplementation with bee pollen and propolis on growth performance and serum metabolites of rabbits: a meta-analysis. Animals (Basel) 2023; 13(3): 439.
[http://dx.doi.org/10.3390/ani13030439] [PMID: 36766327]

[50]   Attia YA, Zeweil HS, Alsaffar AA, El-Shafy AS. Effect of non-antibiotic feed additives as an alternative to flavomycin on broiler chickens' production. Arch Geflugelkd 2011; 75(1): 40-8.

[51]   Gould IM. The epidemiology of antibiotic resistance. Int J Antimicrob Agents 2008; 32 (Suppl. 1): S2-9.
[http://dx.doi.org/10.1016/j.ijantimicag.2008.06.016] [PMID: 18757182]

[52]   Catry B, Laevens H, Devriese LA, Opsomer G, de Kruif A. Antimicrobial resistance in livestock. J

Vet Pharmacol Ther 2003; 26(2): 81-93.
[http://dx.doi.org/10.1046/j.1365-2885.2003.00463.x] [PMID: 12667177]

[53]   van den Bogaard A. Antimicrobial resistance--relation to human and animal exposure to antibiotics. J Antimicrob Chemother 1997; 40(3): 453-4.
[http://dx.doi.org/10.1093/jac/40.3.453] [PMID: 9338504]

[54]   van den Bogaard A, Stobberingh EE. Epidemiology of resistance to antibiotics links between animals and humans. Int J Antimicrob Agents 2000; 14(4): 327-35.
[http://dx.doi.org/10.1016/S0924-8579(00)00145-X] [PMID: 10794955]

[55]   De Leener E. Comparison of antimicrobial resistance among human and animal enterococci with emphasis on the macrolide-lincosamide-streptogramin group 2008.
[http://dx.doi.org/10.1128/AEM.71.5.2766-2770.2005]

[56]   Barton MD. Does the use of antibiotics in animals affect human health? Aust Vet J 1998; 76(3): 177-80.
[http://dx.doi.org/10.1111/j.1751-0813.1998.tb10124.x] [PMID: 9578753]

[57]   Bager F, Aarestrup FM, Madsen M, Wegener HC. Glycopeptide resistance in *Enterococcus faecium* from broilers and pigs following discontinued use of avoparcin. Microb Drug Resist 1999; 5(1): 53-6.
[http://dx.doi.org/10.1089/mdr.1999.5.53] [PMID: 10332722]

[58]   Smith JW II, O'Quinn PR, Goodband RD, Tokach MD, Nelssen JL. Effects of low-protein, amino acid-fortified diets formulated on a net energy basis on growth performance and carcass characteristics of finishing pigs. J Appl Anim Res 1999; 15(1): 1-16.
[http://dx.doi.org/10.1080/09712119.1999.9706225]

[59]   Thompson S, Miller M, Vose D, *et al.* Draft risk assessment on the human health impact of fluoroquinoline resistant Campylobacter associated with consumption of chicken [Web Page]. 2000, Accessed 2010 Oct 12.

[60]   Zakeri A, Bozorgmehri Fard MHV, Feizi A. Effectiveness of 3-Nitro 4- Hydroxy Phenyl Arsenic Acid on Growth – Production Prameters & effectiveness of Coccidioacetat. Iran Magazin of Veterinary Sciences 2005; 2: 3-10.

[61]   Attia Y A, Bovera F, Abd El-Hamid AE, Tag El-Din AE, Al-Harthi MA, El-Shafy AS. Effect of zinc bacitracin and phytase on growth performance, nutrient digestibility, carcass and meat traits of broilers". Journal of Animal Physiology and Animal Nutrition 2016;100:485-491, 2016 in press.
[http://dx.doi.org/10.1111/jpn.12397]

[62]   Windisch W, Schedle K, Plitzner C, Kroismayr A. Use of phytogenic products as feed additives for swine and poultry. J Anim Sci 2008; 86(14) (Suppl. 14): E140-8.
[http://dx.doi.org/10.2527/jas.2007-0459] [PMID: 18073277]

[63]   Kamel C. A novel look at a classic approach of plant extracts. Feed Mix 2000; 11: 19-21.

[64]   Lopez-Bote CJ. Bioflavonoid effects reach beyond productivity. Feed Mix 2004; 12: 12-5.

[65]   Lee Y, Ding P. Physiological production of essential oil in plants - Ontogeny, secretory structures and seasonal variations. RE:view 2016; 2: 1-11.

[66]   Maass N, Bauer J, Paulicks BR, Böhmer BM, Roth-Maier DA. Efficiency of *Echinacea purpurea* on performance and immune status in pigs. J Anim Physiol Anim Nutr (Berl) 2005; 89(7-8): 244-52.
[http://dx.doi.org/10.1111/j.1439-0396.2005.00501.x] [PMID: 15972074]

[67]   Roth-Maier DA, Bohmer BM, Maass N, Damme K, Paulicks BR. Efficiency of *Echinacea purpurea* on performance of broilers and layers. Arch Geflugelkd 2005; 69: 123-7.

[68]   Catala-Gregori P, Mallet S, Travel A, Lessire M. Un extrait de plante et un probiotique sont aussi efficaces que l'avilamycine pour améliorer les performances du poulet de chair. VIIemeJournées de la Recherche Avicole, Tours, France 2007; 202-206.

[69]   Chrubasik S, Pittler MH, Roufogalis BD. *Zingiberis rhizoma*: A comprehensive review on the ginger

effect and efficacy profiles. Phytomedicine 2005; 12(9): 684-701.
[http://dx.doi.org/10.1016/j.phymed.2004.07.009] [PMID: 16194058]

[70]   Platel K, Srinivasan K. Digestive stimulant action of spices: a myth or reality? Indian J Med Res 2004; 119(5): 167-79.
[PMID: 15218978]

[71]   Jamroz D, Wertelecki T, Houszka M, Kamel C. Influence of diet type on the inclusion of plant origin active substances on morphological and histochemical characteristics of the stomach and jejunum walls in chicken. J Anim Physiol Anim Nutr (Berl) 2006; 90(5-6): 255-68.
[http://dx.doi.org/10.1111/j.1439-0396.2005.00603.x] [PMID: 16684147]

[72]   Shen SY, Lin YY, Liao SC, Wang JS, Wang SD, Ching-Yi L. Effects of phytogenic feed additives on the growth, blood biochemistry, and caecal microorganisms of White Roman geese. Czech J Anim Sci 2023; 68(5): 202-11.
[http://dx.doi.org/10.17221/205/2022-CJAS]

[73]   Cuppett SL, Hall CA III. Antioxidant activity of the Labiatae. Adv Food Nutr Res 1998; 42: 245-71.
[http://dx.doi.org/10.1016/S1043-4526(08)60097-2] [PMID: 9597729]

[74]   Placha I, Takacova J, Ryzner M, *et al.* Effect of thyme essential oil and selenium on intestine integrity and antioxidant status of broilers. Br Poult Sci 2014; 55(1): 105-14.
[http://dx.doi.org/10.1080/00071668.2013.873772] [PMID: 24397472]

[75]   Nakatani N. Phenolic antioxidants from herbs and spices. Biofactors 2000; 13(1-4): 141-6.
[http://dx.doi.org/10.1002/biof.5520130123] [PMID: 11237173]

[76]   Wei A, Shibamoto T. Antioxidant activities and volatile constituents of various essential oils. J Agric Food Chem 2007; 55(5): 1737-42.
[http://dx.doi.org/10.1021/jf062959x] [PMID: 17295511]

[77]   Oni AI, Adeleye OO, Adebowale TO, Oke OE. The role of phytogenic feed additives in stress mitigation in broiler chickens. J Anim Physiol Anim Nutr (Berl) 2024; 108(1): 81-98.
[http://dx.doi.org/10.1111/jpn.13869] [PMID: 37587717]

[78]   Sari M, Biondi DM, Kaâbeche M, *et al.* Chemical composition, antimicrobial and antioxidant activities of the essential oil of several populations of Algerian *Origanum glandulosum* Desf. Flavour Frag. J 2006; 21: 890-898.
[http://dx.doi.org/10.1002/ffj.1738]

[79]   Burt, Sara. Essential oils: Their antibacterial properties and potential applications in foods—A review. International Journal of Food Microbiology 2004; 94(3), 223-253. International journal of food microbiology. 94. 223-53.
[http://dx.doi.org/10.1016/j.ijfoodmicro.2004.03.022]

[80]   Newton SM, Lau C, Gurcha SS, Besra GS, Wright CW. The evaluation of forty-three plant species for *in vitro* antimycobacterial activities; isolation of active constituents from *Psoralea corylifolia* and *Sanguinaria canadensis.* J Ethnopharmacol 2002; 79(1): 57-67.
[http://dx.doi.org/10.1016/S0378-8741(01)00350-6] [PMID: 11744296]

[81]   Jamroz D, Orda J, Kamel C, Wiliczkiewicz A, Wertelecki T, Skorupińska J. The influence of phytogenic extracts on performance, nutrient digestibility, carcass characteristics, and gut microbial status in broiler chickens. J Anim Feed Sci 2003; 12(3): 583-96.
[http://dx.doi.org/10.22358/jafs/67752/2003]

[82]   Mitsch P, Zitterl-Eglseer K, Köhler B, Gabler C, Losa R, Zimpernik I. The effect of two different blends of essential oil components on the proliferation of *Clostridium perfringens* in the intestines of broiler chickens. Poult Sci 2004; 83(4): 669-75.
[http://dx.doi.org/10.1093/ps/83.4.669] [PMID: 15109065]

[83]   Botsoglou NA, Christaki E, Fletouris DJ, Florou-Paneri P, Spais AB. The effect of dietary oregano essential oil on lipid oxidation in raw and cooked chicken during refrigerated storage. Meat Sci 2002;

62(2): 259-65.
[http://dx.doi.org/10.1016/S0309-1740(01)00256-X] [PMID: 22061420]

[84]   Ruberto, Giuseppe, Baratta M, Madani, Sarri. Chemical composition and antioxidant activity of essential oils from Algerian *Origanum glandulosum* Desf. Flavour and Fragrance Journal - Flavour Frag J 2002; 17. 251-254.
[http://dx.doi.org/10.1002/ffj.1101]

[85]   Aksit M, Goksoy E, Kok F, Ozdemir D, Ozdogan M. The impacts of organic acid and essential oil supplementations to diets on the microbiological quality of chicken carcasses. Arch Geflugelkd 2006; 70: 168-73.

[86]   Gulmez M, Oral N, Vatansever L. The effect of water extract of sumac (*Rhus coriaria* L.) and lactic acid on decontamination and shelf life of raw broiler wings. Poult Sci 2006; 85(8): 1466-71.
[http://dx.doi.org/10.1093/ps/85.8.1466] [PMID: 16903480]

[87]   Young JF, Stagsted J, Jensen SK, Karlsson AH, Henckel P. Ascorbic acid, alpha-tocopherol, and oregano supplements reduce stress-induced deterioration of chicken meat quality. Poult Sci 2003; 82(8): 1343-51.
[http://dx.doi.org/10.1093/ps/82.8.1343] [PMID: 12943308]

[88]   Govaris A, Florou-Paneri P, Botsoglou E, Giannenas I, Amvrosiadis I, Botsoglou N. The inhibitory potential of feed supplementation with rosemary and/or α-tocopheryl acetate on microbial growth and lipid oxidation of turkey breast during refrigerated storage. Lebensm Wiss Technol 2007; 40(2): 331-7.
[http://dx.doi.org/10.1016/j.lwt.2005.10.006]

[89]   Jambwa P, Nkadimeng SM, Mudimba TN, Matope G, McGaw LJ. Antibacterial and anti-inflammatory activity of plant species used in traditional poultry ethnomedicine in Zimbabwe: A first step to developing alternatives to antibiotic poultry feed additives. J Ethnopharmacol 2023; 300: 115687.
[http://dx.doi.org/10.1016/j.jep.2022.115687] [PMID: 36084819]

CHAPTER 3

Phytobiotics in Animal Nutrition

Mayada R. Farag[1], Mahmoud M. Alagawany[2], Mohamed E. Abd El-Hack[2], Mohammed A. E. Naiel[3], Mahmoud Madkour[4], Abdulmohsen H. Alqhtani[5], Vincenzo Tufarelli[6], Youssef A. Attia[7,8,*], Asmaa F. Khafaga[9] and Maria Cristina de Oliveira[10]

[1] *Forensic Medicine and Toxicology Department, Veterinary Medicine Faculty, Zagazig University, Zagazig-44519, Egypt*

[2] *Poultry Department, Faculty of Agriculture, Zagazig University, Zagazig-44519, Egypt*

[3] *Department of Animal Production, Faculty of Agriculture, Zagazig University, Zagazig-44511, Egypt*

[4] *Animal Production Department, National Research Centre, Giza, Egypt*

[5] *Department of Animal Production, College of Food and Agricultural Sciences, King Saud University, Riyadh, Saudi Arabia*

[6] *Department of Precision and Regenerative Medicine and Jonian Area (DiMePRe-J), Section of Veterinary Science and Animal Production, University of Bari Aldo Moro, 70010 Valenzano, Bari, Italy*

[7] *Sustainable Agriculture Production Research Group, Agriculture Department, Faculty of Environmental Sciences, King Abdulaziz University, Jeddah-21589, Saudi Arabia*

[8] *Animal and Poultry Production Department, Faculty of Agriculture, Damanhour University, Damanhour-22713, Egypt*

[9] *Department of Pathology, Faculty of Veterinary Medicine, Alexandria University, Apis, Alexandria, 21944, Egypt*

[10] *Faculty of Veterinary Medicine, University of Rio Verde, Rio Verde, GO, 75.901-970, Brazil*

Abstract: The modern animal industry faces a persistent challenge: meeting growing consumer demand for high-quality, low-cost food while maintaining stringent standards of sanitation, health, and welfare. In recent decades, antibiotic-supplemented diets have been widely adopted to maximize the growth potential of livestock. However, alternative approaches have emerged, including the use of phytochemicals as substitutes for antibiotics, to enhance avian productivity. Phytobiotics, which consist of herbs and their derivatives, have numerous therapeutic effects and are available in various forms. Recently, this type of feed manipulation has gained popularity in the animal industry as an alternative to antibiotics, primarily because of the lack of adverse

* **Corresponding author Youssef A. Attia:** Sustainable Agriculture Production Research Group, Agriculture Department, Faculty of Environmental Science, King Abdulaziz University, Jeddah-21589, Saudi Arabia & Animal and Poultry Production Department, Faculty of Agriculture, Damanhour University, Damanhour-22516, Egypt; E-mail: yaattia@kau.edu.sa

side effects and their ability to bolster the immune system and improve stress tolerance. In addition to enhancing intestinal integrity and reducing gut damage, phytobiotics promote increased feed intake by compensating for the nutritional demands of local and systemic immune responses. Furthermore, they reduce the concentration of pathogenic microflora in the gastrointestinal tract and mitigate the local inflammatory responses. In poultry, these benefits are demonstrated by improved feed consumption, increased digestive enzyme secretion, and enhanced immune function. Phytobiotics exhibit a wide range of biological activities, including immunity-boosting, antibacterial, antiviral, coccidiostatic, antiparasitic, anti-inflammatory, and antioxidant properties. Herbs and their derivatives have been used since ancient times for their health benefits and minimal side effects. However, recent studies have highlighted that certain herbs and their metabolites may pose risks, raising concerns among consumers regarding the safety of using these compounds as feed supplements or treatments. This chapter explores the beneficial effects and latest developments related to phytobiotics and highlights their practical applications and health advantages. Understanding these features is essential for veterinarians, scientists, pharmacists, physiologists, pharmaceutical industries, nutritionists, and animal breeders as they consider the use of phytobiotics in modern animal husbandry.

Keywords: Essential oils, Herbs, Nutritionists, Phytobiotics, Reproductive performance.

INTRODUCTION

Herbs have emerged as a relatively new class of growth promoters, attracting significant interest in the animal feed industry in recent years. This category includes a wide range of herbs, spices, and their derivatives, with essential oils being the most common [1, 2]. For thousands of years, humans have utilized plant-based products for nutrition and treatment of various ailments [3]. In ancient societies, natural remedies derived from herbs and spices have been used as feed supplements for domestic animals, paralleling the use of pharmaceutical products. Windisch *et al.* [4] defined phytobiotics as plant-derived constituents incorporated into the diets of livestock to enhance performance. This term distinguishes them from plant products used for pharmacological purposes such as the prevention and treatment of specific health conditions. As efforts to eliminate antibiotic growth promoters have intensified in many regions, phytobiotics have emerged as a promising natural alternative containing biologically active compounds. Unlike synthetic antibiotics or inorganic chemicals, phytobiotics are natural, residue-free, and non-toxic [5], making them suitable as feed supplements in livestock production.

This chapter explores the various beneficial applications and newly discovered aspects of phytobiotics. These features are particularly relevant to scientists, physiologists, nutritionists, pharmacists, veterinarians, the pharmaceutical industry, and animal breeders.

PHYTOBIOTICS AND THEIR CHARACTERISTICS

Phytobiotics are derived from various plant parts, including the leaves, roots, flowers, and even entire plants. The products may consist of dried whole plants or plant portions as well as extracts containing specific bioactive components. Phytobiotics are generally classified based on their primary and secondary plant compounds [6]. Primary compounds include basic nutrients such as proteins and fats, while secondary compounds comprise essential (ethereal) and volatile oils, bitters, pungent substances, colorants, and phenolic compounds, among others [7].

In general, phytobiotics have minimal effects on the intake of primary nutrients by animals. Thus, secondary plant compounds are of primary interest because of their biologically active properties [8 - 10].

Phytobiotics can be categorized into four main classes, each of which encompasses a wide array of substances.

- Spices: Plants with strong aromas or flavors commonly used as food additives (*e.g.*, garlic, ginger, and cinnamon).
- Herbs: Non-woody, non-perennial flowering plants, used either whole or in part (*e.g.*, flowers, roots, leaves).
- Essential oils: Volatile lipophilic components extracted from plants.
- Oleoresins: Concentrated extracts from spices containing both volatile and non-volatile compounds [11] (Fig. **1**).

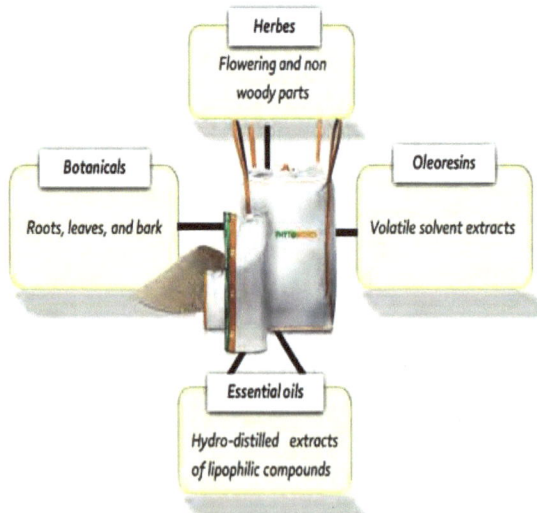

Fig. (1). Classes of phytobiotics.

The active components of phytobiotics are secondary plant metabolites that function as antimicrobial agents and contribute to their therapeutic properties [7].

Numerous factors influence the shelf life of foods, including intrinsic factors such as pH, water activity, nutrient content, presence of antimicrobial agents, redox potential, and physical properties of biological structures, such as temperature [12]. External factors also play a critical role, including storage conditions, relative humidity, and composition of the surrounding environment [13].

When selecting phytobiotics for inclusion in animal diets, four key factors must be considered: i) the specific plant parts used and their physical characteristics, ii) botanical source, iii) timing of harvest, and iv) compatibility of these components with other ingredients in the feed. These factors may account for the variability in the productive outcomes observed when different phytobiotic supplements are incorporated into animal diets. Further research is recommended to elucidate the precise mechanisms of action, dietary compatibility, toxicological profiles, and safety considerations before their broader application in animal feed [10].

Flavonoids are a diverse class of polycyclic phenolic compounds derived from plants [14], and over 4,000 structures have been identified [15, 16], each of which exhibits a range of biological effects in mammals.

Among these, flavonoids may exert estrogenic effects, functioning as nonsteroidal or phytoestrogens. Owing to their structural similarity to estradiol, they can bind to and activate estrogen receptors in cells. However, because of their relatively low binding affinity, they are classified as weak estrogens [18], displaying biological activities between 10^{-2} and 10^{-5} that of 17β-estradiol [19].

Research by McGarvey *et al.* [20] and Dominguez-López *et al.* [21] indicated that phytoestrogens may inhibit gonadotropin-releasing hormone (GnRH) and disrupt the hypothalamic-pituitary-gonadal axis, which regulates estrogen secretion. Helal *et al.* [22] demonstrated these effects in male albino rats, where daily administration of 1 ml/kg body weight of fennel oil for one month led to reduced levels of FSH, testosterone, and sperm count, relative to the control group.

Essential oils may also exhibit hormone-like activity if their molecular structures are sufficiently similar to those of endogenous hormones, allowing them to interact with the same receptors. For instance, anethole, present in the essential oils of star anise, aniseed, and both sweet and bitter fennel, exhibits strong estrogenic activity [23]. Moreover, estrogens play a critical role in regulating lipid metabolism by modulating lipogenesis, lipolysis, and adipogenesis. Phytoestrogens can mimic these effects in adipose tissues [24, 25].

Howes *et al.* [26] proposed that several common essential oil compounds might interact with estrogen receptors, displaying either weak estrogenic or anti-estrogenic activity. Compounds such as citral, geraniol, nerol, and trans-anethole demonstrated estrogenic activity in yeast screens, whereas eugenol exhibited anti-estrogenic properties.

Beneficial Effects of Phytobiotics

The beneficial effects of phytobiotics are shown in Fig. (**2**). Various herbal plants contain bioactive compounds, including phytoestrogens, which possess galactogenic properties, and have been traditionally used as herbal remedies to stimulate milk production in livestock. For example, goat rue (*Galega officinalis*) and shatavari (*Asparagus racemosus*) have been shown to enhance lactation [27 - 30]. Studies have also demonstrated that feeding pigs and goats with a commercial herbal galactagogue formulation containing *A. racemosus* resulted in a significant increase in milk yield and growth of mammary glands, alveolar tissue, and acini [31].

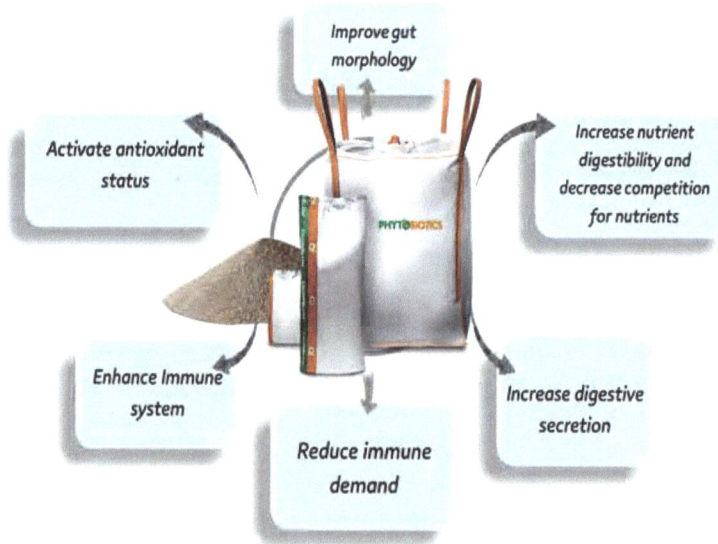

Fig. (2). Beneficial effects of phytobiotics.

Galega officinalis has been shown to improve insulin sensitivity and is believed to be a precursor to the widely used antidiabetic drug metformin [32, 33]. Insulin plays a crucial role in mammary gland function during the early stages of breast development and is essential for colostrum production (lactogenesis I) as well as lactogenesis II, which is characterized by the increased production of milk following the birth of the placenta, typically around the 2nd or 3rd day postpartum, and the continuation of lactation [34]. Furthermore, González-Andrés *et al.* [35] demonstrated that administering low doses of phytoestrogens or diets containing *G. officinalis* could activate specific estrogen receptors in sheep, thereby enhancing milk production. According to Bhaat [36], the active phytochemicals in *G. officinalis* positively influence both the quantity and quality of the milk produced in calves. These active compounds include phenols, flavonoids, saponins, terpenes, and sterols, among others [37].

The phytoestrogens and alkaloids present at high concentrations in *G. officinalis* are also believed to contribute to the anticancer properties of its aqueous extracts [38].

Metformin, originally derived from *Galega officinalis*, has demonstrated antiproliferative effects in tumor cells in preclinical *in vitro* and *in vivo* studies. It acts synergistically with certain anticancer drugs and helps overcome chemoresistance in several types of tumors [32]. Extracts from the aerial parts of *G. officinalis* have shown activity against brain glioblastoma carcinoma and human lung cancer cell lines [39]. Galegine, a naturally occurring compound in *G. officinalis*, has been found to induce cytotoxicity and apoptosis in human melanoma cells [40]. Champavier *et al.* [41] and González-Andrés *et al.* [35] reported that various phytoestrogens, such as flavonol triglycosides, kaempferol, and quercetin, have been isolated from methanol extracts of *G. officinalis*.

The lactogenic potential of *G. officinalis* should be considered when feeding cattle, as it increases both milk yield and lactation persistence [42]. González-Andrés *et al.* [35] confirmed the lactogenic effect in sheep (2 g DM/kg body weight), with a 16.9% increase in total milk yield.

Asparagus racemosus contains steroidal saponins [43, 44], which exhibit a range of pharmacological effects, including cytotoxicity through apoptosis, oncosis, autophagy, and suppression of metastasis; anti-inflammatory properties *via* inhibition of inflammatory mediators; antidiabetic effects through mechanisms such as activation of glycogen synthesis and suppression of gluconeogenesis; and antitumor activity by modulating various signaling pathways. Other effects include hepatoprotection and antifungal activity [45].

Sharma *et al.* [46] proposed that steroidal saponins exert hormone-like effects. Hamed [47] demonstrated that *A. racemosus* has estrogenic effects on the genital organs and mammary glands of rats, causing hyperplasia in alveolar tissues and acini, and enhancing milk production.

According to Pal and Mishra [48], preparations containing *A. racemosus* stimulate hematopoietic function and increase the weight of the accessory sex glands. Additionally, *A. racemosus* has been shown to increase ovarian weight in young females and stimulate folliculogenesis, as evidenced by the histological analysis of the ovaries in immature female rats. The reproductive system, both male and female, relies on glycogen as an estrogen-dependent energy source. Estrogen elevates glycogen levels in the uterus, and any decrease in uterine glycogen can be indicative of estrogen deficiency [49 - 51]. Gopumadhavan *et al.* [52] demonstrated that *A. racemosus* extracts increased uterine weight and glycogen levels when compared to ovariectomized rats.

The pubertal development of mammary glands observed in rabbits treated with aqueous extracts of *A. racemosus* and *G. officinalis* during the rearing period, compared to controls, may be linked to the significant increase in estrogen and progesterone levels in treated does [47]. Furthermore, feeding pigs and goats with a commercial herbal galactagogue formulation containing *A. racemosus* led to a substantial increase in milk supply and the growth of mammary glands, alveolar tissues, and acini [31].

Even at high doses, *A. racemosus* extracts have been reported as safe for long-term use, including during lactation [53, 54]. In a study evaluating the acute and chronic toxicity of *A. racemosus* (1 g/kg) on pre-and post-natal development in rats, it was concluded that the plant is safe, as no differences were observed in overall behavior, locomotion, feed and water consumption, and liver and kidney function between the treatment groups [55].

Female rats were administered the *A. racemosus* extract every other day for 14 days at doses of 300 and 600 mg/kg, with ibuprofen (20 mg/kg) used as a comparative control. The treatment resulted in an increased frequency of the proestrus and estrus phases, along with a reduction in the metaestrus and diestrus phases at both dosage levels. A 100% mating success rate was observed and there was a notable reduction in the force and frequency of uterine contractions. These effects are likely attributable to the estrogenic properties of the plants [56]. However, in a study involving pregnant rats fed with *A. racemosus* root at 10 mg/100 g/day for 60 days, there was a higher incidence of fetal resorption, teratogenic disorders, gross malformations, and intrauterine growth retardation in pups [57]. These findings suggest that *A. racemosus* should be used cautiously during pregnancy [58].

Further research is required to investigate the effects of standardizing procedures, dosage rates, and the duration of administration of these herbal treatments on the metabolic rate of relevant bodily tissues [1, 2, 59].

CONCLUSION AND REMARKS

In conclusion, phytobiotics have emerged as a promising alternative to antibiotic growth promoters in animal husbandry, offering numerous benefits without the associated risks of antibiotic resistance and residues and the emergence of resistant pathogenic strains globally. The diverse range of phytobiotics, including herbs, spices, essential oils, and oleoresins, provide a wealth of bioactive compounds with antimicrobial, anti-inflammatory, and immunomodulatory properties. Notably, plants such as *Galega officinalis* and *Asparagus racemosus* have demonstrated significant potential for enhancing lactation and reproductive performance in livestock. However, while the benefits of phytobiotics are evident, it is crucial to acknowledge that further research is necessary to fully elucidate their mechanisms of action, optimal dosages, and potential long-term effects. As the animal industry continues to evolve, the integration of phytobiotics into feed formulations represents a sustainable and effective approach to improving animal health, productivity, and welfare while addressing consumer demands for natural and safe food products. However, some plants and their metabolites can cause minor side effects. Therefore, their use requires careful consideration and further investigation to ensure their safety and efficacy.

IMPLICATIONS

- The use of phytobiotics could help address the challenge of meeting the growing consumer demand for high-quality, low-cost animal products while maintaining animal health and welfare standards without relying on antibiotics.
- Phytobiotics may help slow down the development of antibiotic-resistant bacteria, which is a major global health concern.
- Phytobiotics offer a more natural approach to enhancing animal health and productivity, potentially appealing to consumers seeking "cleaner" food products.
- The multiple beneficial effects of phytobiotics, including enhanced immunity and stress tolerance, could lead to improved overall animal health and welfare.
- If phytobiotics are effective alternatives to antibiotics, they could help maintain or improve livestock productivity while potentially reducing the costs associated with antibiotic use and the management of antibiotic resistance.
- This review highlights the need for further research on the mechanisms of action, optimal dosages, and potential long-term effects of phytobiotics on animal production.
- As the use of phytobiotics in animal feed increases, regulatory bodies may need to develop new guidelines for safe and effective use.

- The complex nature of phytobiotics and their effects necessitate collaboration among various fields, including veterinary medicine, pharmacology, nutrition, and animal science.
- As the use of phytobiotics becomes more widespread, there may be a need for consumer education regarding these natural feed additives and their benefits.
- Research on phytobiotics could lead to the development of new specialized feed products for the animal industry.

REFERENCES

[1]   Shehata AA, Attia Y, Khafaga AF, *et al.* Restoring healthy gut microbiome in poultry using alternative feed additives with particular attention to phytogenic substances: Challenges and prospects. German Journal of Veterinary Research 2022; 2(3): 32-42.
[http://dx.doi.org/10.51585/gjvr.2022.3.0047]

[2]   Shehata AA, Yalçın S, Latorre JD, *et al.* Probiotics, prebiotics, and phytogenic substances for optimizing gut health in poultry. Microorganisms 2022; 10(2): 395.
[http://dx.doi.org/10.3390/microorganisms10020395] [PMID: 35208851]

[3]   Chaachouay N, Douira A, Zidane L. Herbal medicine used in the treatment of human diseases in the Rif, Northern Morocco. Arab J Sci Eng 2022; 47(1): 131-53.
[http://dx.doi.org/10.1007/s13369-021-05501-1] [PMID: 33842189]

[4]   Windisch W, Schedle K, Plitzner C, Kroismayr A. Use of phytogenic products as feed additives for swine and poultry1. J Anim Sci 2008; 86(14) (Suppl. 14): E140-8.
[http://dx.doi.org/10.2527/jas.2007-0459] [PMID: 18073277]

[5]   Vasilopoulos S, Dokou S, Papadopoulos GA, *et al.* Dietary supplementation with pomegranate and onion aqueous and cyclodextrin encapsulated extracts affects broiler performance parameters, welfare and meat characteristics. Poultry 2022; 1(2): 74-93.
[http://dx.doi.org/10.3390/poultry1020008]

[6]   Ndomou SCH, Mube HK. The use of plants as phytobiotics: a new challenge. In: Soto-Hernández M, Aguirre-Hernández E, Palma-Tenango M, Eds. Phytochemicals in agriculture and food. IntechOpen 2023.
[http://dx.doi.org/10.5772/intechopen.110731]

[7]   Placha I, Gai F, Pogány Simonová M. Editorial: Natural feed additives in animal nutrition—Their potential as functional feed. Front Vet Sci 2022; 9: 1062724.
[http://dx.doi.org/10.3389/fvets.2022.1062724] [PMID: 36439337]

[8]   Attia Y, Al-Harthi M, El-Kelawy M. Utilisation of essential oils as a natural growth promoter for broiler chickens. Ital J Anim Sci 2019; 18(1): 1005-12.
[http://dx.doi.org/10.1080/1828051X.2019.1607574]

[9]   Attia YA, El-Din A, Zeweil HS, Hussein AS, Qota ESM, Arafat MA. The effect of supplementation of enzyme on laying and reproductive performance in Japanese quail hens fed nigella seed meal. J Poult Sci 2008; 45(2): 110-5.
[http://dx.doi.org/10.2141/jpsa.45.110]

[10]   Attia YA, Bakhashwain AA, Bertu NK. Thyme oil (*Thymus vulgaris L.*) as a natural growth promoter for broiler chickens reared under hot climate. Ital J Anim Sci 2017; 16(2): 275-82.
[http://dx.doi.org/10.1080/1828051X.2016.1245594]

[11]   Akosile OA, Kehinde FO, Oni AI, Oke OE. Potential implications of in ovo feeding of phytogenics in poultry production. Transl Anim Sci 2023; 7(1): txad094.
[http://dx.doi.org/10.1093/tas/txad094] [PMID: 37701128]

[12]   Amit SK, Uddin MM, Rahman R, Islam SMR, Khan MS. A review on mechanisms and commercial

aspects of food preservation and processing. Agric Food Secur 2017; 6(1): 51.
[http://dx.doi.org/10.1186/s40066-017-0130-8]

[13]   FSAI – Food Safety Authority of Ireland. Validation of product shelf-life (revision 2) 2014. Available from:
https://www.lenus.ie/bitstream/handle/10147/348584/GN%2018%20Rev%202%20FINAL%20ACCE
SSIBLE.pdf?sequence=1&isAllowed=y

[14]   Rohloff J. Analysis of phenolic and cyclic compounds in plants using derivatization techniques in combination with GC-MS-based metabolite profiling. Molecules 2015; 20(2): 3431-62.
[http://dx.doi.org/10.3390/molecules20023431] [PMID: 25690297]

[15]   Abbas M, Saeed F, Anjum FM, *et al.* Natural polyphenols: An overview. Int J Food Prop 2017; 20(8): 1689-99.
[http://dx.doi.org/10.1080/10942912.2016.1220393]

[16]   Kaurinovic B, Vastag D. Flavonoids and phenolic acids as potential natural antioxidants. In: Shalaby E, Ed. Antioxidants. IntechOpen 2018.
[http://dx.doi.org/10.5772/intechopen.83731]

[17]   D'Arrigo G, Gianquinto E, Rossetti G, Cruciani G, Lorenzetti S, Spyrakis F. Binding of androgen- and estrogen-like flavonoids to their cognate (non)nuclear receptors: a comparison by computational prediction. Molecules 2021; 26(6): 1613.
[http://dx.doi.org/10.3390/molecules26061613] [PMID: 33799482]

[18]   Zhang X, Wu C. *In silico, in vitro* and *in vivo* evaluation of the developmental toxicity, estrogenic activity, and mutagenicity of four natural phenolic flavonoids at low exposure levels. ACS Omega 2022; 7(6): 4757-68.
[http://dx.doi.org/10.1021/acsomega.1c04239] [PMID: 35187296]

[19]   Kummer V, Mašková J, Čanderle J, Zralý Z, Neča J, Machala M. Estrogenic effects of silymarin in ovariectomized rats. Vet Med (Praha) 2001; 46(1): 17-23.
[http://dx.doi.org/10.17221/7846-VETMED]

[20]   McGarvey C, Cates PS, Brooks AN, *et al.* Phytoestrogens and gonadotropin-releasing hormone pulse generator activity and pituitary luteinizing hormone release in the rat. Endocrinology 2001; 142(3): 1202-8.
[http://dx.doi.org/10.1210/endo.142.3.8015] [PMID: 11181536]

[21]   Domínguez-López I, Yago-Aragón M, Salas-Huetos A, Tresserra-Rimbau A, Hurtado-Barroso S. Effects of dietary phytoestrogens on hormones throughout a human lifespan: a review. Nutrients 2020; 12(8): 2456.
[http://dx.doi.org/10.3390/nu12082456] [PMID: 32824177]

[22]   Helal EGE, Abdul-Azis N, El-Aleem MA, Ahmed SS. Effect of phytoestrogen (fennel) on some sex hormones and other physiological parameters in male albino rats. Egypt J Hosp Med 2022.

[23]   Zhang S, Chen X, Devshilt I, *et al.* Fennel main constituent, trans-anethole treatment against LPS-induced acute lung injury by regulation of Th17/Treg function. Mol Med Rep 2018; 18(2): 1369-76.
[http://dx.doi.org/10.3892/mmr.2018.9149] [PMID: 29901094]

[24]   Cederroth CR, Vinciguerra M, Kühne F, *et al.* A phytoestrogen-rich diet increases energy expenditure and decreases adiposity in mice. Environ Health Perspect 2007; 115(10): 1467-73.
[http://dx.doi.org/10.1289/ehp.10413] [PMID: 17938737]

[25]   Kuryłowicz A. Estrogens in adipose tissue physiology and obesity-related dysfunction. Biomedicines 2023; 11(3): 690.
[http://dx.doi.org/10.3390/biomedicines11030690] [PMID: 36979669]

[26]   Howes M-JR, Houghton PJ, Barlow DJ, Pocock VJ, Milligan SR. Assessment of estrogenic activity in some common essential oil constituents. J Pharm Pharmacol 2002; 54(11): 1521-8.
[http://dx.doi.org/10.1211/002235702216] [PMID: 12495555]

[27]   Kumar S, Mehla RK, Singh M. Effect of Shatavari (*Asparagus racemosus*) on milk production and immunemodulation in Karan Fries crossbred cows. Indian J Tradit Knowl 2014; 13: 404-8.

[28]   Saini VP, Choudhary S, Tanwar R, Choudhary SD, Sirvi SP, Yadav VS. Effect of feeding shatavari (*Asparagus racemosus*) root powder on qualitative and quantitative parameter of milk in crossbred cows. Int J Curr Microbiol Appl Sci 2018; 7(8): 3265-77.
[http://dx.doi.org/10.20546/ijcmas.2018.708.348]

[29]   Palka S, Kmiecik M, Migdal L, Siudak Z. The effect of a diet containing fennel (*Foeniculum vulgare* Mill.) and goat's-rue (*Galega officinalis* L.) on litter size and milk yield in rabbits. Scientific Annals of Polish Society of Animal Production 2019; 15: 73-8.
[http://dx.doi.org/10.5604/01.3001.0013.6484]

[30]   Chavan MK, Bhosale TR, Deokar DK. Effect of feeding shatavari (*Asparagus racemosus*) root powder on quantity of milk in crossbred cows. J Dairy Foods Home Sci 2022; 42(Of): 150-3.
[http://dx.doi.org/10.18805/ajdfr.DR-1820]

[31]   Chawla A, Chawla P, Mangalesh RR, Roy RC. *Asparagus racemosus* (Willd): biological activities & its active principles. Indo Global Journal of Pharmaceutical Sciences 2011; 1(2): 113-20.
[http://dx.doi.org/10.35652/IGJPS.2011.11]

[32]   Christodoulou MI, Scorilas A. Metformin and anti-cancer therapeutics: hopes for a more enhanced armamentarium against human neoplasias? Curr Med Chem 2017; 24(1): 14-56.
[http://dx.doi.org/10.2174/0929867323666160907161459] [PMID: 27604091]

[33]   Stuebe AM. Medical complications of mothers. 2022. 9th ed. p. 546-571.
[http://dx.doi.org/10.1016/B978-0-323-68013-4.00015-8]

[34]   Berlato C, Doppler W. Selective response to insulin *versus* insulin-like growth factor-I and -II and up-regulation of insulin receptor splice variant B in the differentiated mouse mammary epithelium. Endocrinology 2009; 150(6): 2924-33.
[http://dx.doi.org/10.1210/en.2008-0668] [PMID: 19246539]

[35]   González-Andrés F, Redondo PA, Pescador R, Urbano B. Management of *Galega officinalis* L. and preliminary results on its potential for milk production improvement in sheep. N Z J Agric Res 2004; 47(2): 233-45.
[http://dx.doi.org/10.1080/00288233.2004.9513591]

[36]   Bhatt N. Herbs andd herbal supplements, a novel nutritional approach in animal nutrition. Iran J Appl Anim Sci 2015; 497-515.

[37]   Attia YA, Bakhashwain AA, Bertu NK. Utilisation of thyme powder (*Thymus vulgaris* L.) as a growth promoter alternative to antibiotics for broiler chickens raised in a hot climate. European Poultry Science 2018; 82: eps.2018.238.
[http://dx.doi.org/10.1399/eps.2018.238]

[38]   Yildirim AB, Karakas FP, Turker AU. *In vitro* antibacterial and antitumor activities of some medicinal plant extracts, growing in Turkey. Asian Pac J Trop Med 2013; 6(8): 616-24.
[http://dx.doi.org/10.1016/S1995-7645(13)60106-6] [PMID: 23790332]

[39]   Karakas FP, Turker AU, Karakas A, Mshvildadze V. Cytotoxic, anti-inflammatory and antioxidant activities of four different extracts of *Galega officinalis* L. (goat's rue). Trop J Pharm Res 2016; 15(4): 751-7.
[http://dx.doi.org/10.4314/tjpr.v15i4.12]

[40]   Arjmand MH, Sabri H, Maghrouni A, Zarei E, Hotelchi M, Afshari AR. Potential cytotoxic activity of galegine on human melanoma cells. Research Journal of Pharmacognosy 2022; 9: 15-23.
[http://dx.doi.org/10.22127/rjp.2022.324923.1829]

[41]   Champavier Y, Allais DP, Chulia AJ, Kaouadji M. Acetylated and non-acetylated flavonol triglycosides from *Galega officinalis*. Chem Pharm Bull (Tokyo) 2000; 48(2): 281-2.
[http://dx.doi.org/10.1248/cpb.48.281] [PMID: 10705519]

[42] Penagos Tabares F, Bedoya Jaramillo JV, Ruiz-Cortés ZT. Pharmacological overview of galactogogues. Vet Med Int 2014; 2014: 1-20.
[http://dx.doi.org/10.1155/2014/602894] [PMID: 25254141]

[43] Saxena S. Levels of diabetic retinopathy. In: Saxena S, Ed. Jaypee gold standard mini atlas series: diabetic retinopathy. Jaypee Brothers Medical Publishers Ltd. 2012; pp. 15-62.
[http://dx.doi.org/10.5005/jp/books/11558_3]

[44] Saxena G, Singh M, Bhatnagar M. Phytoestrogens of *Asparagus racemosus* Wild. Journal of Herbal Medicine and Toxicology 2010; 4: 15-20.

[45] Porte S, Joshi V, Shah K, Chauhan NS. Plants' steroidal saponins – a review on its pharmacology properties and analytical techniques. World J Tradit Chin Med 2022; 8(3): 350-85.
[http://dx.doi.org/10.4103/2311-8571.353503]

[46] Sharma P, Chauhan PS, Dutt P, *et al.* A unique immuno-stimulant steroidal sapogenin acid from the roots of *Asparagus racemosus.* Steroids 2011; 76(4): 358-64.
[http://dx.doi.org/10.1016/j.steroids.2010.12.006] [PMID: 21172369]

[47] Hamed R. Effect of aqueous extracts of *Galega officinalis* and *Asparagus racemosus* supplementation on development of mammary gland, milk yield and its impact on the productivity of rabbit does. Egypt Poult Sci 2016; 36(4): 985-1004.
[http://dx.doi.org/10.21608/epsj.2016.168817]

[48] Pal RS, Mishra A. Dhatryadi ghrita: medicine for complete biological care for women. World Journal of Environmental Biosciences 2020; 9: 37-44.

[49] Dean M. Glycogen in the uterus and fallopian tubes is an important source of glucose during early pregnancy. Biol Reprod 2019; 101(2): 297-305.
[http://dx.doi.org/10.1093/biolre/ioz102] [PMID: 31201425]

[50] Sandoval K, Berg MD, Guadagnin AR, Cardoso FC, Dean M. Endometrial glycogen metabolism on days 1 and 11 of the reproductive cycle in dairy cows. Anim Reprod Sci 2021; 233: 106827.
[http://dx.doi.org/10.1016/j.anireprosci.2021.106827] [PMID: 34450335]

[51] Silva R, Carrageta DF, Alves MG, Oliveira PF. Testicular glycogen metabolism: an overlooked source of energy for spermatogenesis? BioChem 2022; 2(3): 198-214.
[http://dx.doi.org/10.3390/biochem2030014]

[52] Gopumadhavan S, Venkataranganna MV, Rafiq M, Madhumathi BG, Mitra SK. Evaluation of the estrogenic effect of menosan using the rat models of uterotrophic assay. Medicine Update 2005; 13: 37-41.

[53] Alok S, Jain SK, Verma A, Kumar M, Mahor A, Sabharwal M. Plant profile, phytochemistry and pharmacology of *Asparagus racemosus* (Shatavari): A review. Asian Pac J Trop Dis 2013; 3(3): 242-51.
[http://dx.doi.org/10.1016/S2222-1808(13)60049-3]

[54] Kushwah P, Ghulaxe SPC, Mandloi N, Singh S, Patel R, Patel R. Review on medicinal value of *Asparagus racemosus* in woman's. Research Journal of Pharmacy and Technology 2018; 11(1): 418-20.
[http://dx.doi.org/10.5958/0974-360X.2018.00077.X]

[55] Kim JC, Shin DH, Kim SH, *et al.* Peri- and postnatal developmental toxicity of the fluoroquinolone antibacterial DW-116 in rats. Food Chem Toxicol 2004; 42(3): 389-95.
[http://dx.doi.org/10.1016/j.fct.2003.10.002] [PMID: 14871581]

[56] Kaaria LM, Oduma JA, Kaingu CK, Mutai PC, Wafula DK. Effect of *Asparagus racemosus* on selected female reproductive parameters using Wistar rat model. Discovery Phytomedicine 2019; 6(4): 199-204.
[http://dx.doi.org/10.15562/phytomedicine.2019.110]

[57]   Goel RK, Prabha T, Kumar MM, Dorababu M, Prakash , Singh G. Teratogenicity of *Asparagus racemosus* Willd. root, a herbal medicine. Indian J Exp Biol 2006; 44(7): 570-3. [https://www.mdpi.com/2072-6643/12/8/2456#]. [PMID: 16872047]

[58]   McGuire TM. Drugs affecting milk supply during lactation. Aust Prescr 2018; 41(1): 7-9. [http://dx.doi.org/10.18773/austprescr.2018.002] [PMID: 29507453]

[59]   Gálik B, Wilkanowska A, Bíro D, *et al.* Effect of a phytogenic additive on blood serum indicator levels and fatty acids profile in fattening turkeys meat. J Cent Eur Agric 2015; 16: 383-98. [http://dx.doi.org/10.5513/JCEA01/16.4.1642]

CHAPTER 4

Thyme

Mohamed E. Abd El-Hack[1], Mahmoud M. Alagawany[1], Youssef A. Attia[2,3,*], Adel D. Al-qurashi[3], Abdulmohsen H. Alqhtani[4], Bahaa Abou-Shehema[5], Ayman E. Taha[6], Ahmed A. Abdallah[5], Mohamed A. AlBanoby[7], Nehal K. Bertu[8], Vincenzo Tufarelli[9] and Omer H.M. Ibrahim[3,10]

[1] *Poultry Department, Faculty of Agriculture, Zagazig University, Zagazig-44519, Egypt*

[2] *Animal and Poultry Production Department, Faculty of Agriculture, Damanhour University, Damanhour-22713, Egypt*

[3] *Sustainable Agriculture Production Research Group, Agriculture Department, Faculty of Environmental Sciences, King Abdulaziz University, Jeddah-21589, Saudi Arabia*

[4] *Department of Animal Production, College of Food and Agricultural Sciences, King Saud University, Riyadh, Saudi Arabia*

[5] *Department of Poultry Nutrition, Animal Production Institute, Agricultural Research Center, Dokki, Giza-3751310, Egypt*

[6] *Department of Animal Husbandry and Animal Wealth Development, Faculty of Veterinary Medicine, Alexandria University, Apis, Alexandria, 21944, Egypt*

[7] *Al-Shamel Animal Feed Factory, Industrial Area, Hail-55411, Saudi Arabia*

[8] *Directorate of Agriculture, Animal Production Department, Beheira Governorate, Damanhour, Egypt*

[9] *Department of Precision and Regenerative Medicine and Jonian Area (DiMePRe-J), Section of Veterinary Science and Animal Production, University of Bari Aldo Moro, 70010 Valenzano, Bari, Italy*

[10] *Department of Floriculture, Faculty of Agriculture, Assiut University, Egypt*

Abstract: *Thymus vulgaris* L., a member of the *Lamiaceae* family, is a herb widely used in conventional medicine because of its various therapeutic properties. Thyme, mostly cultivated in the Mediterranean region, is used as a spice and medicine worldwide, owing to its antioxidant and antibacterial properties. This chapter focuses on data supporting the use of thyme as a productive enhancer in animal feed and as a partial or full substitute for antibiotics. The essential oils found in the aerial portions of thyme are a source of fragrance and therapeutic qualities. The main active constituents

* **Corresponding authors Youssef A. Attia and Mohamed E. Abd El-Hack:** Animal and Poultry Production Department, Faculty of Agriculture, Damanhour University, Damanhour-22713, Egypt, Sustainable Agriculture Production Research Group, Agriculture Department, Faculty of Environmental Sciences, King Abdulaziz University, Jeddah-21589, Saudi Arabia and Department of Poultry, Faculty of Agriculture, Zagazig University, Zagazig-44519, Egypt; E-mail: dr.mohamed.e.abdalhaq@gmail.com

of thyme extract are thymol, carvacrol, and other mono- and sesquiterpenes. These compounds contribute to the flavor, fragrance, and antibacterial properties of thyme. The effect of thyme on animal performance is attributed to its bioactive compounds, which vary depending on several factors. Thyme supplementation has been documented to be advantageous in poultry production, with thymol reported to prevent oral bacterial infections and to influence the permeability of pathogenic bacterial cell walls, leading to cell death. Essential oils from thyme can also support digestive functions by stimulating endogenous enzyme activity, nitrogen absorption, and regulating the ammonia content and odor of excreta. The antibacterial properties of thyme are influenced by the chemical structure and lipophilic characteristics of its essential oils, allowing them to pass through bacterial membranes and affect the interior of the cell.

Keywords: Antibiotic, Feed additives, Growth promoters, Poultry, Thyme.

INTRODUCTION

Thyme (*Thymus vulgaris* L.) (Fig. **1**), a member of the *Lamiaceae* family and regionally known as "zaatar" or "zaitra" in Arabic, is used extensively in folk medicine for its expectorant, antitussive, antispasmodic, antibroncholitic, anthelmintic, carminative, and diuretic properties. Thyme is a widely used medicinal herb that is mostly cultivated in the Mediterranean region and used worldwide for culinary and medicinal purposes. Owing to its antibacterial and antioxidant properties, it is a herbal plant that has received increasing attention [1 - 3]. Many fragrant herbs of significant scientific and commercial value, including thyme, marjoram, mint, rosemary, and oregano belong to the *Lamiaceae* family. The essential oils found in the peltate glandular trichomes in the aerial portions of the herbs are the source of the scent associated with this vegetation. Highly specialized secretory cells are found in glandular trichomes, where essential oils are produced and stored in a subcuticular storage chamber [4]. *Thyme* is a widespread plant worldwide because of its fragrant and therapeutic qualities. Some *Thymus* species are utilized as medicinal herbs, herbal teas, and flavorings (condiments and spices) [5]. The genus *Thymus* includes 18 wild species and one cultivated species, *Thymus vulgaris* L., a fragrant plant [6]. *Thymus vulgaris* exhibits polymorphic diversity in the synthesis of monoterpenes, and the genus *Thymus* is known for its intraspecific chemotype variation. The names of the six chemotypes were derived from their predominant monoterpenes: α-terpineol (A), geraniol (G), thuyanol-4 (U), carvacrol (C), linalool (L), and thymol (T) [5, 7, 8]. (Fig. **1**) shows the primary active ingredients found in the thyme extract.

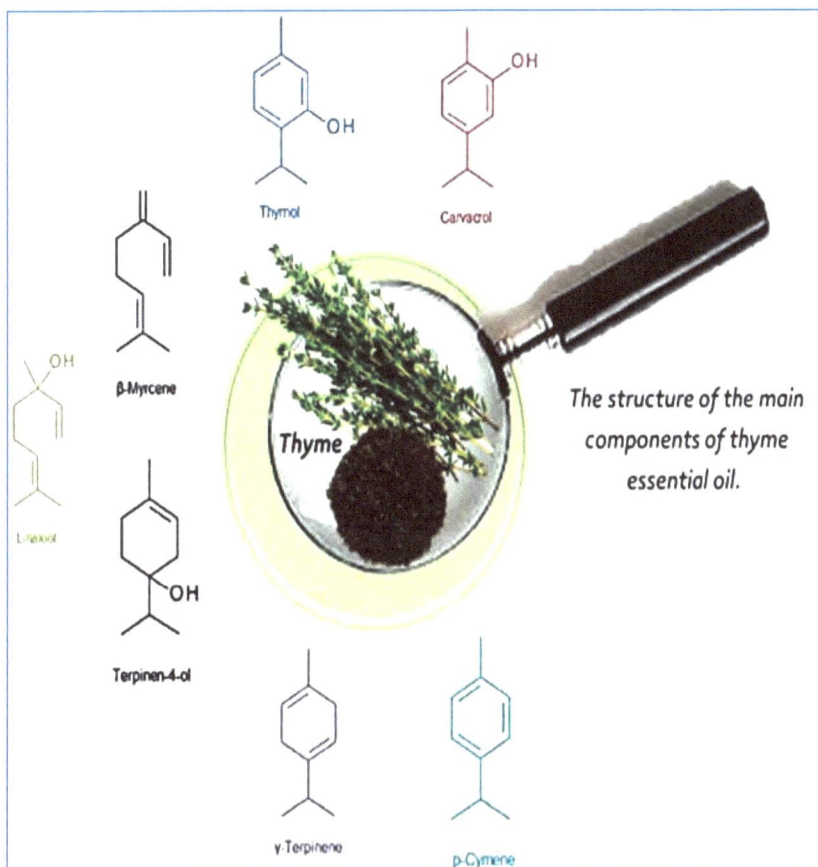

Fig. (1). The main active constituents present in thyme extract.

Thymus vulgaris contains several active compounds, including phenols, such as thymol (40%) and carvacrol (15%), according to Mikaili *et al.* [1]. Furthermore, Stahl-Biskup [5] discovered that mono- and sesquiterpenes predominate in the makeup of marjoram and thyme essential oils. These substances give these plants their flavor and fragrance, and the essential oils extracted can be utilized as antibacterial or antiseptic agents in pharmaceutical and medical applications, as well as in the production of cosmetics and fragrances [5, 9, 10]. Mono- and sesquiterpenes are believed to protect vegetation from herbivores and infections [11]. Carvacrol and thymol, two monoterpenes of the *Lamiaceae* family frequently found in thyme, have attracted considerable attention. The main benefits of these two phenolic monoterpenes are their antiherbivorous, antibacterial, and antioxidant properties [12 - 15]. A total of 41 components (97.85% of the total detectable compounds) were identified by GC and GC–MS analysis of the essential oil from thyme [5]. β-pinene (4.32%); borneol (5.03%), 1, 8-cineole (5.57%), α-pinene (9.55%), camphene (17.57%), and camphor (39.39%)

were the constituents of the oil in ascending order. Less than 2% of the other components were present in the thyme essential oil. The most significant monoterpene hydrocarbons, camphene, α-pinene, β-pinene, and myrcene, were found in the thyme essential oil (Table **1**). Oxygenated monoterpenes accounted for 54.82% of the total oil content, and the chief components of thyme essential oil were thymol and carvacrol. Thymol (5-methyl- 1-2-isopropyl phenol) and carvacrol (5-isopropyl-2-methyl phenol) are the principal phenolic constituents of *Thymus vulgaris* [16].

Table 1. Volatile components identified in thyme samples [2, 3, 5, 16, 17].

Compound	Thyme Volatile Oil (%)	Thyme Liposoluble Fraction (%)
Monoterpene hydrocarbon		
α-pinene	1.23	-
Camphene	0.63	-
β-pinene	0.32	-
Myrcene	1.63	-
α-terpinene	0.8	-
P-cymene	30.53	0.81
Limonene	0.62	-
Sabinene	4.24	-
Monoterpene esters		
1,8-cineole	1.24	-
Monoterpene alcohols		
Linalool	2.73	-
α-terpineol	1.24	-
Geraniol	0.64	-
Borneol	3.16	-
Monoterpene ketones		
Camphor	0.83	-
Monoterpenic esters		
Bornyl acetate	0.7	-
Terpenoidic phenols		
Thymol	30.86	1.01
Carvacrol	3.37	-
Sesquiterpene hydrocarbons		
Caryophyllene	2.48	-

MODE OF ACTION OF THYME

The effect of phytogenic plants on animal performance is attributed to their bioactive compounds [14, 15] and not their influence on energy and protein intake [18 - 20]. Depending on the plant type, individual plant parts, and agronomic and environmental factors, these chemicals can vary significantly [3, 21]. Commercial antibiotics can be replaced with thyme plants. The advantages of thyme in the poultry sector have been widely documented [22]. Thymol has been reported to prevent oral bacteria infections [23]. Moreover, studies [24 - 26] documented the advantages of thyme supplementation in chicken production. According to Lee *et al.* [27], thymol can influence the permeability of the cell walls of pathogenic bacteria, causing hole development, osmotic stress, and release of the cytoplasm and its active components outside the cell, ultimately resulting in cell death. The bactericidal activity of thymol on these bacteria was described by Bolukbasi and Erhan [28] as a result of the disruption of the essential membrane ions of potassium and hydrogen equilibrium pumps. Essential oils have been primarily used to manage infections through their antioxidant and antibacterial properties in the past, and are now able to support digestive processes by stimulating endogenous enzyme activity, nitrogen absorption, and regulating ammonia content and odor of the excreta [29, 30]. By interfering with the cell membrane, thymol alters the permeability of cations such as K^+ and H^+, thereby exerting antibacterial effects [31]. The antibacterial properties of plant extracts, particularly volatile oils, have been the subject of several investigations [10, 32 - 34]. According to some theories, the antibacterial action of essential oils may be influenced by their chemical structures and lipophilic characteristics [35, 36]. Owing to their lipophilicity, terpenoids and phenylpropanoids may be able to pass through bacterial membranes and affect the inside of the cell [37]. Furthermore, the antibacterial action of essential oils is also a result of structural characteristics, such as aromaticity [38] and the presence of functional groups [35]. Because poultry may not respond to flavor like pigs, the specific effects of essential oils on chicken performance have not been studied extensively [39]. Nonetheless, some studies [40] suggest that taste may have a little impact on hen feed consumption. The mechanism of action of thyme in fowl gut is briefly described in Fig. (**2**).

Antioxidant Property of Thyme

Thyme appears to affect *in vivo* antioxidant defense mechanisms, such as glutathione peroxidase, vitamin E, and superoxide dismutase. In addition, it acts as a potent free radical scavenger [18]. Based on the decreased quantity of malonaldehyde in the yolk, there is evidence that the antioxidant components of thyme can be transmitted to egg yolks through hen nutrition [41]. The antioxidant qualities of thyme are due to both thymol and D-cymene-2, 3-dio [17]. As phenolic components have antioxidant and free radical-scavenging properties, they are thought to support optimal health [42].

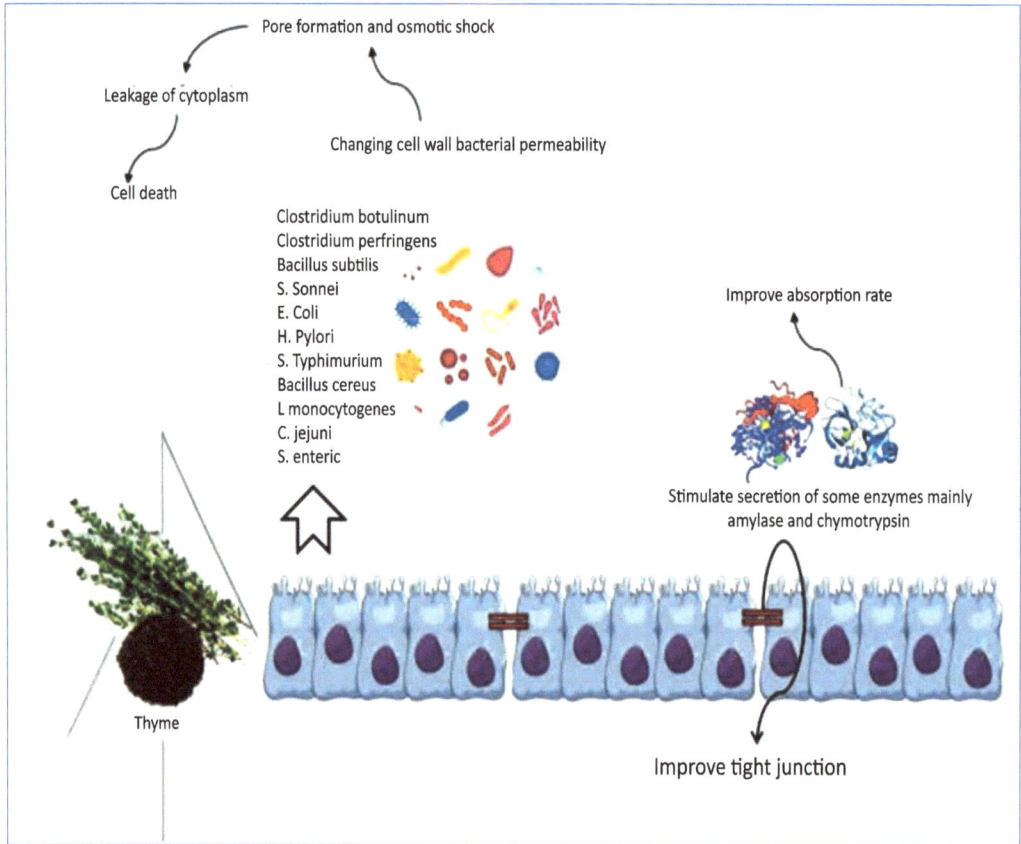

Fig. (2). The mode of action of thyme in poultry intestinal cells.

Antimicrobial Properties of Thyme

Studies have shown that thyme or its main polyphenolic constituents have antibacterial activity against *C. perfringens*, *C. botulinum*, *S. sonnei*, *Bacillus subtilis*, *H. pylori*, *E. coli*, *Bacillus cereus*, *S. typhimurium*, *C. jejuni*, *L monocytogenes*, and *S. enteric* [25, 43 - 50]. The influence of thyme extract, essential oils, powder, and other major constituents on poultry performance has been demonstrated [51 - 53]. When added to the medium (*in vitro*), ground thyme has been shown to inhibit *S. typhimurium* development [54]. Moreover, it has been demonstrated that thyme essential oil inhibits *in vitro* the development of *E. coli* in medium [55]. Thyme oil demonstrated its antibacterial action against *C. perfringens*, *S. epidermis*, and *S. serovars* only at high doses (500 mg/l); whilst at low levels, it can be active (50 mg/l) against *E. coli* [56]. Additionally, thyme has antibacterial properties against *S. enteritidis*, *S. aureus*, *B. serovar* [33], *S. morium* [50], *E. coli* [32], *C. perfringens* A [57, 58], and *S. pneumonia* [57]. Approximately 20-55% of thyme volatile oil extract is composed mostly of thymol. Its primary characteristics are linked to these oils, which have been investigated in light of their antibacterial properties [58]. Similar to thymol, carvacrol also exhibits antibacterial activity. Carvacrol and thymol have controllable values of 100–1000 ppm, based on laboratory antibacterial investigations. The antibacterial activity of carvacrol and thymol against *C. jejuni*, *S. enterica*, *E. Coli*, and *L. monocytogenes* was recently validated *in vitro*, and cinnamaldehyde, carvacrol, and thymol were found to be the most active against *E. coli* by Friedman *et al.* [48].

Mechanical Effect of Thyme Oil on Digestion

The digestive system of broilers, particularly the intestine, is affected by thyme oil extracts, which results in the release of endogenous digestive enzymes. Amylase and chymotrypsin are two primary enzymes whose synthesis is increased by the thyme extract. This leads to an increase in intestinal absorption rate, which in turn increases the rate of eating. As a result, they lose more weight and the feed conversion ratio (FCR) decreases. In this manner, there will be greater and more economical body weight following slaughter. Additionally, the weight of viscera, such as the liver and gizzard, increases because of volatile thyme oils [2, 10, 31, 50, 60, 61].

Effect of Carvacrol on Digestive Enzymes

An herbaceous perennial plant called nettle contains carvacrol, which has stimulatory effects on pancreatic secretions [62]. This is achieved by increasing the secretion of digestive enzymes, which increases the amount of nutrients, such as amino acids, that can be absorbed and digested from the digestive tract and enhance the characteristics of the carcass. By physically grinding and increasing bile production during nutritional digestion, nettle can increase the percentage of gizzards and the liver. Organs such as the thigh and breast expand more in response to higher absorption levels of amino acids [63]. Because thymol and carvacrol from thyme extract have antibacterial properties, they help eliminate pathogens from broiler intestines, improving health, growth, and, more generally, production. According to several studies, thyme oil can help, regardless of intestinal and digestive issues [32, 58]. Age-related variations exist in the positive response of the broiler digestive system, with younger animals showing stronger reactions and, therefore, better weight growth [2, 3, 14, 33]. The benefits of thyme for gut health [15] and the digestive system [61] are amplified when combined with soybean meal.

Preference Use of Thyme Oil in Animal Feed

Since thyme oil is an herbal product, it does not have negative effects on chemical growth boosters, including drug resistance, deposition in animal flesh, increased risk of illness after usage, or high costs [33]. Bacterial cell membranes appear to be the primary targets of essential oils [21]. As a result, gram-positive bacteria are often more sensitized to the bactericidal action of thyme oil than gram-negative bacteria [64]. Numerous studies have documented the beneficial effects of herbal plants on broilers [62, 65]. These investigations highlight the antifungal, antioxidant, and anticoccidial properties of herbs. Secondary metabolites of herbs, including phenols, essential oils, and saponins, have been linked to negative medicinal effects [2, 3, 10, 14, 15, 18, 66]. Consequently, it was shown that adding thyme essential oil to the diet had a limited impact on performance metrics and enhanced the microbiological conditions of meat [67].

POULTRY PERFORMANCE

Kalantar *et al.* [68] showed that there were substantial variations in feed consumed, gain, and feed efficiency across the groups; hence, the thyme essence treatments had better means than the control group. The control group had the lowest feed intake and growth, whereas the 0.2% thyme essence group had the highest level. The lowest energy intake was relative to the average daily gain and maximum gain per protein consumption (pertained to the 0.2% thyme essence level). There was no significant difference in the energy/protein ratio among treatments. According to Toghyani *et al.* [69], broiler body weight (BW) and feed utilization (FCR) were significantly influenced by thyme at a low dose (5 g/kg) but not at a high dose (10 g/kg). Furthermore, BW and FCR improved in the group fed thyme, according to Najafi and Torki [70]. Additionally, Griggs and Jacob [71] proposed that thyme is effective against bacterial infections in poultry. According to Abd El-Hakim *et al.* [72], thyme is essential for development up to day 21 when compared to the control, citric acid, and lactic acid groups. Furthermore, Feizi *et al.* [73] demonstrated that thyme oil influenced broiler feed intake, body weight gain (BWG), and FCR. Throughout the trial, broilers fed the control diet had a lower FCR and greater BWG than those fed the thyme diet (especially the 1000 ppm group). In addition, the BWG was greater in the thyme diet group than in the control group. These findings suggest that 1000 ppm thyme was more effective than 500 ppm thyme in promoting growth. Furthermore, compared to the control group, which performed the worst, Al-Kassie [50] observed that chicks fed 200 ppm thyme essential oils had considerably greater feed intake, BWG, and FCR. Al-Tebary [74] demonstrated that the thyme group performed much better than the control group; nevertheless, there were no appreciable variations in FCR between the two groups. Additionally, the death rate dropped in the thyme group, but it was not significantly different from that in the other groups. Essential oils have recently been added to broiler feed to boost growth. Broiler development performance was reported to be stimulated by supplementation with extracted thyme oil; however, studies on essential oils have shown inconsistent outcomes [50, 75, 76]. According to Lee *et al.* [51, 77], supplementing broiler diets with thyme oil increased BWG, feed consumption, and FCR. These improvements may have been caused by active chemicals (thymol and carvacrol), which are thought to stimulate digestion, in addition to their antibacterial action against gut bacteria. Consistent with these findings, carvacrol and a combination of essential oils from cinnamon, oregano, capsicum, and thyme significantly improved the FCR [78]. However, researchers found that feed utilization of broilers was not considerably affected by thyme essential oil [22, 69, 79]. Feizi *et al.* [73] demonstrated that there were notable variations across the groups, with chicks fed 1000 ppm thyme extract exhibiting considerably reduced feed intake, FCR, and mortality rate. Chicks fed 500 ppm thyme extract showed the lowest performance compared to the control group. The 1000 ppm thyme group had the highest BW, which was substantially higher than those of the other two groups. According to Toghyani *et al.* [69], broiler performance was significantly affected by varying thyme dosages (5–10 g/kg) with respect to feed intake, BWG, and FCR. When Al-Kassie [50] examined the effects of thyme on broilers and discovered that it improved poultry health, BWG, feed intake, and FCR. In their investigation of the effects of thyme (1.5%) and garlic (1.5%) on the performance of broiler chicks, Mansoub and Nezhady [65] found that both supplements outperformed the control group at the end of the respective experimental periods. At the conclusion of the breeding period, Feizi *et al.* [80] observed a significant improvement in the BWG and FCR of chickens administered 20% thyme extract compared to the control group. Studies have shown that volatile oils have either beneficial or neutral effects on chick development performance. The amount of thyme added to the diets ranged from 20 to 200 parts per million in different experiments. Performance was positively impacted by volatile oils; feed intake and BWG increased, but FCR was lower than that in the control group [27]. The thyme and mint extracts of broilers were shown to be meaningful by Ocak *et al.* [76], who found that the thyme group was 1.24% heavier than the control group. According to Ayoola *et al.* [81], broiler internal offal, primal section weight, and growth performance are unaffected by dietary thyme leaves. The meat cooking output and loss of the analyzed broilers showed no significant differences. According to this finding, thyme leaves are not a growth booster. However, no adverse effects were observed in these animals. According to Attia *et al.* [2], broiler body weight was not substantially affected by thyme oil at 1, 1.5, or 2 g/kg diet. Thyme essential oil have also demonstrated comparable effects [10]. According to recent research, broiler chickens fed 100 mg/kg of EOs in heat-stable, encapsulated forms might provide an alternative growth booster for zinc broiler chickens [10]. After reviewing the latest research, Shehata *et al.* [15] concluded that the impact of bioactive phytogenic compounds on intestinal health is responsible for the improvements in animal performance.

Feed Intake and Feed Conversion Ratio

Feed intake is influenced by a wide range of factors. Feed selection is influenced by several factors, including appearance, temperature, viscosity, salivation, nutritional value, toxicity of feed components, particle size, and social interactions [82]. There have also been reports of notable increases in feed consumption when 100 or 200 mg/kg of thyme essential oil was used. Compared with the control diet, the application of 0.5% thyme extract greatly increased feed intake, as demonstrated by Bolukbasi *et al.* [52] and Rafiee *et al.* [83]. Nevertheless, when significant amounts (5 g/kg) of thyme essential oil were administered, Cross *et al.* [75] reported a decrease in feed consumption between 8 and 14%. However, later in the test, this impact began to fade. Scientists speculate that this could be due to the strong flavor of thyme essential oil, which the newborn chicks may not have found appetizing at first, but eventually become accustomed to. The predominant flavoring of thyme and oregano essential oils is mostly attributed to carvacrol and its isomer thymol. Lee *et al.* [77] supplemented a broiler meal with 200 mg/kg of carvacrol and thymol to evaluate their individual effects. These findings showed that carvacrol reduced feed intake, which may have been caused by lower hunger in the chicks. In subsequent investigations, Lee *et al.* [27] supplemented diets demonstrating growth-reducing effects with 100 mg/kg thymol or cinnamonaldehyde, either as a result of carboxymethyl cellulose supply or as a result of substituting rye cereal for maize grain. According to Attia *et al.*, thyme oil at 1 g/kg diet dramatically reduced feed consumption compared with the other treatment groups [2]. However, compared with the other thyme groups and the control and thyme oil groups, the same dose of thyme enhanced the FCR and production index. When added to the diet at 10, 20, and 30 g/kg, thyme powder supplementation increased the feed conversion above that of the control [3]. Attia *et al.* [10] discovered that, with the exception of the group given EOs_100, the group that received EOs_150 exhibited considerably better gains (BWG) and feed utilization in comparison to the other treatments. The group that received more than 25 mg/kg of essential oils showed enhanced development, which may have been due to the increased digestibility of lipids, proteins, and fiber.

Mortality Rate

Essential oil supplementation in chicken diets lowered mortality rates according to Bolukbasi *et al.* [52]. Feizi *et al.* [73] revealed that the 1000 ppm thyme group had the lowest death rate whereas the control group had the highest. According to Attia *et al.* [2], the groups fed 1.0, 1.5, and 2.0 g/kg diet of thyme oil, MOS, and the control had comparable survival rates. Similar findings were noted by Attia *et al.* [3, 10] and Shehatat *et al.* [14, 15], suggesting that the beneficial effects of phytogenic plants on animal health were due to improved gut health rather than other effects of bioactive compounds. According to Gumus *et al.*, dietary thyme essential oil significantly reduced the total number of mesophilic aerobic bacteria in drumstick-saved meat and on days 0 and 8 in stored breast meat [67]. There was variation in the microbial counts at different times, as well as in the drumstick and breast-flesh samples. The pH of meat from drumsticks and breasts is unaffected by dietary supplements containing thyme essential oil [67]. The influence of dietary thyme essential oil on water activity and color characteristics of drumstick and breast meat throughout the souring process was negligible and erratic. After 0, 2, and 4 days of storage, thyme essential oil significantly (p<0.01) reduced the levels of the reactive component thiobarbituric acid, a biomarker of drumstick flesh lipid peroxidation. Consequently, it was shown that adding thyme essential oil to the diet had a limited effect on the microbiological condition of meat [67].

Carcass Characteristics

A chicken body that has been killed for meat is the standard description of a chicken corpse [84]. The examination of carcasses is a crucial aspect of poultry processing, as it entails assessing the standards and caliber of the chicks based on their grading. However, several variables, including stress, nutrition, pre-slaughter care, and genetics, can alter the makeup of carcasses [85]. The capacity of poultry production to increase the percentage of carcass and breast meat while lowering fat levels is the main determinant of profitability [86]. According to Rafiee *et al.* [83], broilers fed 0.5% thyme extract have a markedly lower liver percentage. The heart % did not significantly differ between the experimental groups. There is statistical evidence that 0.5% thyme extract reduces the proportion of belly fat. The proportion of drumsticks increased in broilers treated with 0.5% thyme extract. The 0.5% thyme extract resulted in a larger proportion of breast meat than the other extracts; however, no discernible differences were observed. According to the data, the gizzard percentage was lowest in the control groups and greater in the thyme 0.5% extract groups. *Bursa Fabricius* weight percentage was the lowest in the control groups and increased in the thyme 0.5% extract groups. However, when broilers were fed 0.5% thyme extract, which was higher than the control group, the amount of belly fat decreased. Mansoub and Nezhady [65] discovered that the carcass features of broilers were strongly impacted by varying concentrations of thyme powder (0.75, 1, 1.5, and 2%). Group 2 had the highest proportion of breast tissue (35.26%). According to Bolukbasi *et al.*, essential oil supplementation in chicken diets increases weight gain and improves carcass quality [52]. According to Ayoola *et al.* [81], broiler internal offals, primal section weights, and growth performance were unaffected by dietary thyme leaves. The cooking yield and cooking loss of the tested broiler meat did not exhibit significant variation, indicating that thyme leaf is not a growth booster. Compared to the control and Zn bacitracin (antibiotic) groups, the gizzard percentage increased significantly as a result of thyme powder supplementation; the greatest impact was observed at 30 g/kg thyme powder [3]. The proportion of broiler dressings and edible sections of carcasses increased with the addition of essential oils [10].

Economic Efficiency

The largest economic metrics, such as point spread, performance index, and output index, were shown to be related to the levels of 0.2 and 0.15% thyme essence supplementation in chicken diets in a study conducted by Kalantar *et al.* [68]. In conclusion, feeding broiler chicks thyme essence at concentrations of 0.2 and 0.15% may enhance their production performance, energy and protein efficiency, and economic indices. According to Shabaan [86], adding a combination of 0.15% *Cuminum cyminum* and 0.15% thyme to a low-energy diet increased its economic efficiency by 5.79% compared with the control diet. The European production efficiency index of the ZnB group was significantly lower than that of the control group supplemented with 10 and 30 g/kg ZnB [3]. Attia *et al.* [10] stated that the group that got EOs_150 exhibited a considerably enhanced production index compared to the other treatments, except for treatment provided with EOs_100.

Blood Parameters

Rafiee *et al.* [83] revealed that the calcium and phosphorus levels were increased when chicks were given a thyme 0.5% extract diet ($p<0.05$). At 42 days, the antibody titer against NDV was considerably higher in the thyme 0.5% extract group ($p<0.05$). Dietary thyme significantly decreased the mean values of serum constituents in broiler chickens fed different diets, as demonstrated by the serum concentrations of triglycerides and total cholesterol compared to the control group, according to the findings of Mansoub and Nezhady [65]. Thymes contain compounds, such as carvacrol and thymol, which are the primary causes of the drop in blood triglycerides and cholesterol in chicks. These compounds influence blood triglyceride and cholesterol levels, thereby lowering dangerous levels [50, 88]. These effects on cholesterol content are in contrast to earlier research [89] that did not find thyme to have a lowering effect. The level of low-density lipoprotein (LDL) and high-density lipoprotein (HDL) remained unchanged [69,

70]. However, high doses of thyme (10 g/kg) increased HDL content. The examined thyme oil (0.5 g/kg) resulted in a lower malondialdehyde (MDA) level in the kidney and duodenum mucosa, according to Placha *et al.* [90]. Chicks with a lower heterophil-to-lymphocyte ratio drank water at 500 and 1000 parts per million thyme. These findings suggest that heterophil counts and the heterophil-to-lymphocyte ratio increased in other groups in response to an increase in the total bacterial count [73]. Najafi and Torki [70] observed that the number of heterophils was considerably lower in thyme-fed chicks than that in the control group. According to Toghyani *et al.* [69], the heterophil-to-lymphocyte ratio was lower in chicks fed a 5 g/kg thyme diet than that in the control and antibiotic groups. According to Al-Tebary [74], thyme powder considerably decreases variations in blood serum concentrations of glucose and cholesterol, while increasing those of total protein and globulin. However, the levels of albumin, calcium, and phosphorus in the blood serum did not vary markedly between the two groups, though. Feeding thyme extract to broilers resulted in considerably increased antibody titers against NDV, as demonstrated by Rafiee *et al.* [83]. However, according to Mansoub and Nezhady [65], broiler immune responses are unaffected by several doses of thyme powder (0.75 0.1, 1.5, and 2%). Feizi and Nazeri [91] reported fewer reactions in the test group than in the control group. Therefore, thyme essence prevents the onset of the vaccine response. Abdulkarimi *et al.* [43] discovered that using thyme extract at days 21 and 42 of the experiment had no effect on the relative weight of organs, such as the spleen and bursa Fabricius. At 21 days of age, greater bronchitis antibody titers were seen in chicks ingesting 0.2 and 0.6% thyme extract (p<0.05) than in chicks consuming 0.4% of *Thymus vulgaris* extract and the control chicks. Furthermore, in orthogonal comparisons, the bronchitis antibody titer increased in chicks administered *Thymus vulgaris* extract in water compared to the control group. Although there was no impact on the immune response, there were no appreciable differences between the groups in the NDV titer at 21 and 42 days of age and the antibody titer for bronchitis at 42 days of age. According to Zadeh *et al.* [92], there was no significant difference seen in immunological variables, such as bursa and spleen percent and antibody outcomes to sheep red blood cells and NDV, between the chicks given 0.1% *Thymus vulgaris* extract and the control group. When comparing broilers fed 5 and 10 g/kg thyme powder to control chicks, none of the immune-related measures, including antibody titers against NDV, sheep red blood cells, influenza viruses, and albumin to globulin and heterophil to lymphocyte ratios, changed substantially [69]. In addition, dietary thyme extract (0.1%) soluble in water improved production traits and lactic acid counts, while lowering *E. Coli* counts; however, it had no effect on the immune system when compared to the control group, according to Abdulkarimi *et al.* [43]. The amount and type of thyme, time spent in preparation, and immunization regimen employed are likely factors in the absence of immune system effects from thyme extract [59, 89]. According to Attia *et al.* [2], broiler hens given 1.5 and 2.0 g/kg of MOS had more plasma total protein than the control group. The globulin value was greater in the group given 1.5 g/kg of thyme oil than in the control group. Compared with the control group, the albumin-to-globulin ratio was lower in thyme oil at 1.5 g/kg. Compared to the TO group, the control and MOS groups had higher blood AST levels. Thyme supplementation did not influence the RBC properties of red blood cells. Furthermore, the WBC count was higher in the thyme oil 1 group than in the other thyme oil groups. There were no differences in the percentages of monocytes, basophils, eosinophils, or heterophils found in broiler blood across the experimental groups. Compared to the control group, the MOS and TO 2.0 groups had greater blood antibody titers for IBD. There were no discernible variations in the heterophil/lymphocyte (H/L) ratio or serum MDA levels. Compared to the other groups, the addition of 20 and 30 g/kg DTP significantly increased the plasma total protein levels. The mean cell hemoglobin and hemoglobin content were greater in the DTP-10, DTP-20, and ZnB groups than in the control group [3]. White blood cell concentrations were greater in the ZnB group than in the DTP-20 and control groups [3]. Compared to the other groups, the plasma malondialdehyde concentration of the chickens in DTP-10 was comparatively lower. In contrast, DTP-30 had greater plasma total antioxidant capacity than the control group. Furthermore, compared to the control group, the dietary thyme powder-10 group had a noticeably increased antibody titer against infectious bursa disease [3]. The authors [3] came to the conclusion that supplementing broilers raised in a hot climate for one to twenty-eight days with 10 g/kg DTP (11.4 mg/kg thymol and 1.05 mg/kg carvacrol) could be used in place of antibiotic production simulators to enhance FCR and antibody to infectious bursa disease. Significantly elevated plasma levels of globulin and total protein were observed in diets supplemented with EOs_100 and EOs_150; however, plasma glucose concentrations were elevated by EOs_150 [10].

CONCLUSION AND REMARKS

Thymes have emerged as promising natural alternatives to antibiotic growth promoters in poultry production. Growing concerns over antibiotic resistance and residues have driven research into plant-based substitutes that can enhance productivity without compromising animal or human health. Over the past decade, medicinal plants such as thyme have gained increasing attention in the poultry industry. Available evidence indicates that thyme can improve productive performance in birds through multiple mechanisms, including positive effects on the intestinal microbiota and overall health. Its antimicrobial, antioxidant, and growth-promoting properties make it a suitable candidate to replace antibiotics in chicken feed. While more research is needed, the current results on thyme supplementation could allow poultry producers to maintain productivity while addressing concerns regarding antibiotic use. Overall, thyme shows significant potential as a natural feed additive to support sustainable poultry production.

IMPLICATIONS

As antibiotics are growth promoters in chicken production and may have unintended consequences for animal and human health, scientists are searching for viable alternatives to reduce or eliminate their usage. These substitutes have been used more frequently in poultry over the past decade. Medicinal plants, leaves, and flowers have been reported to increase chicken productivity in various ways. Extensive research on thyme as a natural feed additive for poultry has led to significant applications in sustainable poultry production. The antimicrobial, antioxidant, and growth-promoting properties of thyme suggest that it could serve as a viable alternative to antibiotic growth promoters, addressing concerns regarding antibiotic resistance and residues in poultry products. The positive effects of thyme on intestinal health, feed conversion, and overall bird performance indicate its potential to maintain or even enhance productivity while avoiding routine antibiotic use. However, the variability in results across different studies, particularly regarding the optimal dosage and specific effects, implies that further research is needed to standardize the recommendations for thyme supplementation in poultry diets. As the poultry industry continues to seek sustainable and consumer-acceptable production methods, these findings suggest that thyme could play a key role in developing antibiotic-free production systems that meet both productivity and health standards in conventional and organic poultry farming. This could lead to significant changes in feed formulation practices and poultry management strategies in the future. Based on available research, thyme may be classified as a medicinal herb and could serve as a suitable replacement for antibiotics in chicken production, as it can improve the productive performance of birds and, at the same time, have a positive impact on animal health by affecting the intestinal microbiota.

REFERENCES

[1]   Mikaili P, Mohammad Nezhady MA, Shayegh J, Asghari MH. Study of antinociceptive effect of *Thymus vulgaris* and *Foeniculum vulgare* essential oil in mouse. Int J Acad Res 2010; 2010: 2. [Part II.].

[2]   Attia YA, Bakhashwain AA, Bertu NK. Thyme oil (*Thyme vulgaris L.*) as a natural growth promoter for broiler chickens reared under hot climate. Ital J Anim Sci 2017; 16(2): 275-82.
        [http://dx.doi.org/10.1080/1828051X.2016.1245594]

[3]   Attia YA, Bakhashwain A A, Bertu Nehal K. Utilisation of thyme powder (*Thymus vulgaris* L.) as a growth promoter alternative to antibiotics for broiler chickens raised in a hot climate. Europ. Poult. Sci.2018; 82. 2018, 15.
        [http://dx.doi.org/10.1399/eps.2018.238]

[4]   Turner G, Gershenzon J, Nielson EE, Froehlich JE, Croteau R. Limonene synthase, the enzyme responsible for monoterpene biosynthesis in peppermint, is localized to leucoplasts of oil gland secretory cells. Plant Physiol 1999; 120(3): 879-86.
        [http://dx.doi.org/10.1104/pp.120.3.879] [PMID: 10398724]

[5]   Stahl-Biskup E. Essential oil chemistry of the genus *Thymus* - a global view. In: Stahl-Biskup E, Saez F, Eds. Thyme: The genus *Thymus*. 1st ed. London, New York: Taylor and Francis 2022; Vol. 24: pp. 75-124.

[6]   Marculescu A, Vlase L, Hanganu D, Dragulescu C, Antonie I, Neli-Kinga O. Polyphenols analyses from *Thymus* species, Proc. Rom. Acad, Series B 2007; 3: 117-21.

[7]   Thompson JD, Chalchat JC, Michet A, Linhart YB, Ehlers B. Qualitative and quantitative variation in monoterpene co-occurrence and composition in the essential oil of *Thymus vulgaris* chemotypes. J Chem Ecol 2003; 29(4): 859-80.
        [http://dx.doi.org/10.1023/A:1022927615442] [PMID: 12775148]

[8]   Abd El-Hack ME, El-Saadony MT, Saad AM, *et al*. Essential oils and their nanoemulsions as green alternatives to antibiotics in poultry nutrition: a comprehensive review. Poult Sci 2022; 101(2): 101584.
        [http://dx.doi.org/10.1016/j.psj.2021.101584] [PMID: 34942519]

[9]   Kintzios SE. Profile of the multifaced prince of the herbs. In: Kintzios SE, Ed. Oregano: the genera *Origanum* and *Lippia*. 1st ed. London: Taylor and Francis 2002; 25: pp. 3-8.
        [http://dx.doi.org/10.1201/b12591]

[10]   Attia YA, Mohammed A. Al-Harthi, Mahmoud I El-Kelway. Utilization of essential oils as natural growth promoter for broiler chickens. Ita. J Anim Sci 2019; 18(1): 1005-12.

[11]   Gershenzon J, Dudareva N. The function of terpene natural products in the natural world. Nat Chem Biol 2007; 3(7): 408-14.
        [http://dx.doi.org/10.1038/nchembio.2007.5] [PMID: 17576428]

[12]   Sedy KA, Koschier EH. Bioactivity of carvacrol and thymol against *Frankliniella occidentalis* and *Thrips tabaci*. J Appl Entomol 2003; 127(6): 313-6.
        [http://dx.doi.org/10.1046/j.1439-0418.2003.00767.x]

[13]   Braga PC, Culici M, Alfieri M, Dal Sasso M. Thymol inhibits *Candida albicans* biofilm formation and mature biofilm. Int J Antimicrob Agents 2008; 31(5): 472-7.
        [http://dx.doi.org/10.1016/j.ijantimicag.2007.12.013] [PMID: 18329858]

[14]   Shehata AA, Attia Y, Khafaga AF, *et al*. Restoring healthy gut microbiome in poultry using alternative feed additives with particular attention to phytogenic substances: Challenges and prospects. German Journal of Veterinary Research 2022; 2(3): 32-42.
        [http://dx.doi.org/10.51585/gjvr.2022.3.0047]

[15]   Shehata AA, Yalçın S, Latorre JD, *et al*. Probiotics, prebiotics, and phytogenic substances for optimizing gut health in poultry. Microorganisms 2022; 10(2): 395.
        [http://dx.doi.org/10.3390/microorganisms10020395] [PMID: 35208851]

[16]   Masada Y. Analysis of essential oils by gas chromatography and mass spectrometry. John Wiley and Sons Inc, New York NY1976; pp. 334.0.

[17]   Krause EL, Ternes W. Bioavailability of the antioxidative thyme compounds thymol and p-cymene-2, 3-diol in eggs. Eur Food Res Technol 1999; 209(2): 140-4.
        [http://dx.doi.org/10.1007/s002170050473]

[18]   Attia YA, Alagawany MM, Farag MR, *et al*. Phytogenic products and phytochemicals as a candidate strategy to improve tolerance to COVID-19. Front Vet Sci 2020; 7: 573159.
        [http://dx.doi.org/10.3389/fvets.2020.573159] [PMID: 33195565]

[19]   Attia YA, Al-Harthi MA. *Nigella* seed oil as an alternative to antibiotic growth promoters for broiler

Chickens. Eur Polit Sci 2015; 79: 2015. [DOI: 10.1399/eps.2015.80.].

[20]   Attia YA, El-Din AELRET, Zeweil HS, Hussein AS, Qota ESM, Arafat MA. The effect of supplementation of enzyme on laying and reproductive performance in Japanese quail hens fed nigella seed meal. J Poult Sci 2008; 45(2): 110-5.
[http://dx.doi.org/10.2141/jpsa.45.110]

[21]   Burt S. Essential oils: their antibacterial properties and potential applications in foods—a review. Int J Food Microbiol 2004; 94(3): 223-53.
[http://dx.doi.org/10.1016/j.ijfoodmicro.2004.03.022] [PMID: 15246235]

[22]   Demir E, Kiline K, Yildirim Y, Dincer F, Eseceli H. Comparative effects of mint, stage, thyme and flavomycin in wheat-based broiler diets. Arch Zootech 2008; 11: 54-3.

[23]   Twetman S, Petersson LG. Effect of different chlorhexidine varnish regimens on mutans streptococci levels in interdental plaque and saliva. Caries Res 1997; 31(3): 189-93.
[http://dx.doi.org/10.1159/000262397] [PMID: 9165189]

[24]   Allen PC, Danforth HD, Augustine PC. Dietary modulation of avian coccidiosis. Int J Parasitol 1998; 28(7): 1131-40.
[http://dx.doi.org/10.1016/S0020-7519(98)00029-0] [PMID: 9724884]

[25]   Cross DE, McDevitt RM, Hillman K, Acamovic T. The effect of herbs and their associated essential oils on performance, dietary digestibility and gut microflora in chickens from 7 to 28 days of age. Br Poult Sci 2007; 48(4): 496-506.
[http://dx.doi.org/10.1080/00071660701463221] [PMID: 17701503]

[26]   Mi D, Okan F, Uluocak AN. Effect of dietary supplementation of herb essential oils on the growth performance, carcass and intestinal characteristics of quail. S Afr J Anim Sci 2004; 34: 174-9.

[27]   Lee KW, Ewerts H, Beynen AC. Essentiel oils in broiler nutrition. Int J Poult Sci 2004; 3: 738-52.
[http://dx.doi.org/10.3923/ijps.2004.738.752]

[28]   Bolukbasi S, Erhan M. Effect of dietary thyme and rosemary on laying hens performance and *Escherichia coli* concentration in feces. International Journal of Natural and Engineering Sciences 2007; 1: 55-8.

[29]   Varel VH. Livestock manure odor abatement with plant-derived oils and nitrogen conservation with urease inhibitors: A review1. J Anim Sci 2002; 80(E-suppl_2): E1-7.
[http://dx.doi.org/10.2527/animalsci2002.80E-Suppl_2E1x]

[30]   Azarfar S, Nobakht A, Memannavaz Y, Safamehr AR, Aghdam-Shahryar H. Influence of dietary supplemented thyme (*Thymus vulgaris* L.) and pennyroyal (*Mentha pulegium*) leaves on hematological indices of Japanese quails (*Coturnix coturnix* japonica). Eur J Exp Biol 2012; 2(3): 605-7.

[31]   Alcicek A, Bozkurt M, Cabuk M. The effect of a mixture of herbal essential oils, an organic acid or a probiotic on broiler performance. S Afr J Anim Sci 2004; 34: 217-22.

[32]   Burt SA, Reinders RD. Antibacterial activity of selected plant essential oils against *Escherichia coli* O157:H7. Lett Appl Microbiol 2003; 36(3): 162-7.
[http://dx.doi.org/10.1046/j.1472-765X.2003.01285.x] [PMID: 12581376]

[33]   Dalkilic B, Guler T, Ertas OR, Ciftci M. III. National Animal Nutrition Congress. Adana-Turkey. 2005; pp. 378-82.

[34]   Barbour EK, El-Hakim RG, Kaadi MS, Shaib HA, Gerges DD, Nehme PA. Evaluation of the histopathology of the respiratory system in essential oil-treated broilers following a challenge with *Mycoplasma gallisepticum* and/or H9N2 influenza virus. Int J Appl Res Vet Med 2006; 4: 293-300.

[35]   Farag RS, Daw ZY, Hewedi FM, El-Baroty GSA. Antimicrobial activity of some Egyptian spice essential oils. J Food Prot 1989; 52(9): 665-7.
[http://dx.doi.org/10.4315/0362-028X-52.9.665] [PMID: 31003289]

[36]   Cornner DE, Davidson PM, Al-Branen (Ed.) Antimicrobials in Foods. Marcel Dekker, New York

1993; pp. 441–468.

[37] Helander IM, Alakomi HL, Latva-Kala K, *et al.* Characterization of the action of selected essential oil components on Gram-negative bacteria. J Agric Food Chem 1998; 46(9): 3590-5.
[http://dx.doi.org/10.1021/jf980154m]

[38] Bowles BL, Miller AJ. Antibotulinal properties of selected aromatic and aliphatic aldehydes. J Food Prot 1993; 56(9): 788-94.
[http://dx.doi.org/10.4315/0362-028X-56.9.788] [PMID: 31113058]

[39] Moran ET, Jr. Effect of pellet quality on the performance of meat birds. Pages 87–108 in Recent Advances in Animal Nutrition. Butterworths, London, UK.1989.
[http://dx.doi.org/10.1016/B978-0-408-04149-2.50009-X]

[40] Deyoe CW, Davies RE, Krishnan R, Khaund R, Couch JR. Studies on the taste preference of the chick. Poult Sci 1962; 41(3): 781-4.
[http://dx.doi.org/10.3382/ps.0410781]

[41] Botsoglou NA, Yannakopoulos AL, Fletouris DJ, Tserveni-Goussi AS, Fortomaris PD. Effect of dietary Thyme on the oxidative stability of egg yolk. J Agric Food Chem 1997; 45(10): 3711-6.
[http://dx.doi.org/10.1021/jf9703009]

[42] Ündeğer Ü, Başaran A, Degen GH, Başaran N. Antioxidant activities of major thyme ingredients and lack of (oxidative) DNA damage in V79 Chinese hamster lung fibroblast cells at low levels of carvacrol and thymol. Food Chem Toxicol 2009; 47(8): 2037-43.
[http://dx.doi.org/10.1016/j.fct.2009.05.020] [PMID: 19477215]

[43] Abdulkarimi R, Daneshyar M, Aghazadeh A. Thyme (*Thymus vulgaris*) extract consumption darkens liver, lowers blood cholesterol, proportional liver and abdominal fat weights in broiler chickens. Ital J Anim Sci 2011; 10(2): e20.
[http://dx.doi.org/10.4081/ijas.2011.e20]

[44] Juven BJ, Kanner J, Schved F, Weisslowicz H. Factors that interact with the antibacterial action of thyme essential oil and its active constituents. J Appl Bacteriol 1994; 76(6): 626-31.
[http://dx.doi.org/10.1111/j.1365-2672.1994.tb01661.x] [PMID: 8027009]

[45] Tabak M, Armon R, Potasman I, Neeman I. In *vitro* inhibition of *Helicobacter pylori* by extracts of thyme. J Appl Bacteriol 1996; 80(6): 667-72.
[http://dx.doi.org/10.1111/j.1365-2672.1996.tb03272.x] [PMID: 8698668]

[46] Ultee A, Slump RA, Steging G, Smid EJ. Antimicrobial activity of carvacrol toward *Bacillus cereus* on rice. J Food Prot 2000; 63(5): 620-4.
[http://dx.doi.org/10.4315/0362-028X-63.5.620] [PMID: 10826719]

[47] Fan M, Chen J. Studies on antimicrobial activity of extracts from thymeWei. Sheng Wu Xue Bao 2001; 41:499-504.

[48] Friedman M, Henika PR, Mandrell RE. Bactericidal activities of plant essential oils and some of their isolated constituents against *Campylobacter jejuni, Escherichia coli, Listeria monocytogenes*, and *Salmonella enterica.* J Food Prot 2002; 65(10): 1545-60.
[http://dx.doi.org/10.4315/0362-028X-65.10.1545] [PMID: 12380738]

[49] Nevas M, Korhonen AR, Lindström M, Turkki P, Korkeala H. Antibacterial efficiency of Finnish spice essential oils against pathogenic and spoilage bacteria. J Food Prot 2004; 67(1): 199-202.
[http://dx.doi.org/10.4315/0362-028X-67.1.199] [PMID: 14717375]

[50] Al-Kassie GAM. Influence of two plant extracts derived from Thyme and Cinnamon on broiler performance. Pak Vet J 2009; 29: 169-73.

[51] Lee KW, Everts H, Kappert HJ, Frehner M, Losa R, Beynen AC. Effect of dietary essential oils on growth performance, digestive enzymes and lipid metabolism in female broiler chickens. Br. Poult. Sci. Br Poult Sci 2003; 44: 450-7.
[http://dx.doi.org/10.1080/0007166031000085508] [PMID: 12964629]

[52]   Bolukbasi Ş, Erhan M, Özkan A. Effect of dietary thyme oil and vitamin E on growth, lipid oxidation, meat fatty acid composition and serum lipoproteins of broilers. African Journal of Animal Science 2006; 36: 189-96.

[53]   Al-Mashhadani EH, Farah K, Al-Jaff YM. Farhan, AL-Mashhadani HE. Egyptian Poultry Science J 2011; 31: 481-9.

[54]   Aktug SE, Karapinar M. Sensitivity of some common food-poising bacteria to thyme, mint and bay leaves. Int J of Food 1986; 3: 349-54.

[55]   Marino M, Bersani C, Comi G. Antimicrobial activity of the essential oils of *Thymus vulgaris* L. measured using a bioimpedometric method. J Food Prot 1999; 62(9): 1017-23.
[http://dx.doi.org/10.4315/0362-028X-62.9.1017] [PMID: 10492476]

[56]   Ouwehand AC, Tiihonen K, Kettunen H, Peuranen S, Schulze H, Rautonen N. *In vitro* effects of essential oils on potential pathogens and beneficial members of the normal microbiota. Vet Med (Praha) 2010; 55(2): 71-8.
[http://dx.doi.org/10.17221/152/2009-VETMED]

[57]   Inouye S, Takizawa T, Yamaguchi H. Influence of two plant extracts derived from thyme and cinnamon on broiler performance journal of antimicrobial chemotherapy. 2001; 47: 565-73.

[58]   Losa R, Kohler B. 13th European Symposium on poultry nutrition 2001; 30 Sept-04 Oct 2001 Blankenberge, Belgium.

[59]   Basílico MZ, Basílico JC. Inhibitory effects of some spice essential oils on *Aspergillus ochraceus* NRRL 3174 growth and ochratoxin A production. Lett Appl Microbiol 1999; 29(4): 238-41.
[http://dx.doi.org/10.1046/j.1365-2672.1999.00621.x] [PMID: 10583751]

[60]   Botsoglou NA, Florou-Paneri P, Christaki E, Fletouris DJ, Spais AB. Effect of dietary oregano essential oil on performance of chickens and on iron-induced lipid oxidation of breast, thigh and abdominal fat tissues. Br Poult Sci 2002; 43(2): 223-30.
[http://dx.doi.org/10.1080/00071660120121436] [PMID: 12047086]

[61]   Hernández F, Madrid J, García V, Orengo J, Megías MD. Influence of two plant extracts on broilers performance, digestibility, and digestive organ size. Poult Sci 2004; 83(2): 169-74.
[http://dx.doi.org/10.1093/ps/83.2.169] [PMID: 14979566]

[62]   Mansoub NH. Effect of probiotic bacteria utilization on serum cholesterol and triglycrides contents and performance of broiler chickens. Glob Vet 2010; 5: 184-6.

[63]   Modiry A, Nobakht A, Mehmannavaz Y. Investigation the effects using different mixtures of nettle (*Urtica dioica*), menta pulagum (*Oreganum valgare*) and zizaphora (*Thymyus valgaris*) on performance and carcass traits of broilers. Proc 4th Ir Cong Anim Sci 252-4.

[64]   Kalemba D, Kunicka A. Antibacterial and antifungal properties of essential oils. Current Medicinal Chemistry and Industry. 2003; 10, p. 813-829.
[http://dx.doi.org/10.2174/0929867033457719]

[65]   Mansoub NH, Nezhady MAM. The effect of using thyme, garlic, nettle on performance, carcass quality and blood parameters. Ann Biol Res 2011; 2: 315-20.

[66]   Ipu MA, Akhtar MS, Anjumi MI, Raja ML. New dimension of medicinal plant. Pak Vet J 2006; 26: 144-8.

[67]   Gumus R, Gelen SU. Effects of dietary thyme and rosemary essential oils on performance parameters with lipid oxidation, water activity, pH, colour and microbial quality of breast and drumstick meats in broiler chickens. Arch Tierzucht 2023; 66(1): 17-29.
[http://dx.doi.org/10.5194/aab-66-17-2023] [PMID: 36687212]

[68]   Kalantar M, Saki AA, Zamani P, Aliarabi H. Effect of drinking thyme essence on performance, energy and protein efficiency and economical indices of broiler chickens. Anim Sci J 2011; 24: 59-67.

[69] Toghyani MM, Tohidi M, Gheisari A, Tabeidi SA. Performance, immunity, serum biochemical and hematological parameters in broiler chicks fed dietary thyme as alternative for an antibiotic growth promoter. Afr J Biotechnol 2010; 9: 6819-25.

[70] Najafi P, Torki M. Performance, blood metabolites and immunocompetence of broiler chicks fed diets included essentioal oils of medicinal herbs. J Anim Vet Adv 2010; 9(7): 1164-8.
[http://dx.doi.org/10.3923/javaa.2010.1164.1168]

[71] Griggs JP, Jacob JP. Alternatives to antibiotics for organic poultry production. J Appl Poult Res 2005; 14(4): 750-6.
[http://dx.doi.org/10.1093/japr/14.4.750]

[72] El-Hakim ASA, Cherian G, Ali MN. Use of organic acid, herbs and their combination to improve the utilization of commercial low protein broiler diets. Int J Poult Sci 2008; 8(1): 14-20.
[http://dx.doi.org/10.3923/ijps.2009.14.20]

[73] Feizi A, Bijanza P, Asfaram H, *et al.* Effect of thyme extract on hematological factors and performance of broiler chickens. Pelagia Research Library. Eur J Exp Biol 2014; 4(1): 125-8.

[74] Al-Tebary AS. The effect of dietary *Thymus vulgaris* on some productive and dextran oligosaccharide supplemented diets. International Journal of Poultry Science 2012; 7 (10): 969-977, 2008 and Infectious Diseases, 8, 311-318.

[75] Cross DE, Sovboda K, McDevitt RM, Acamovic T. The performance of chickens fed diets with or without thyme oil and enzymes. Br Poult Sci 2003; 48: 496-506.
[http://dx.doi.org/10.1080/00071660701463221] [PMID: 17701503]

[76] Ocak N, Erener G, Burak Ak F, Sungu M, Altop A, Ozmen A. Performance of broilers fed diets supplemented with dry peppermint (*Mentha piperita* L.) or thyme (*Thymus vulgaris* L.) leaves as growth promoter source. Czech J Anim Sci 2008; 53(4): 169-75.
[http://dx.doi.org/10.17221/373-CJAS]

[77] Lee KW, Everts H, Kapperst HJ, Yeom KH, Beynen AC, Beynen AC. Dietary Carvacrol lowers body weight gain but improves feed conversion in broiler chickens. J Appl Poult Res 2003; 12(4): 394-9.
[http://dx.doi.org/10.1093/japr/12.4.394]

[78] Zhang AW, Lee BD, Lee SK, *et al.* Effects of yeast (*Saccharomyces cerevisiae*) cell components on growth performance, meat quality, and ileal mucosa development of broiler chicks. Poult Sci 2005; 84(7): 1015-21.
[http://dx.doi.org/10.1093/ps/84.7.1015] [PMID: 16050118]

[79] Tekeli A, Celik L, Kutlu HR, Gorgulu M. Effect of dietary supplemental plant extracts on performance, carcass characteristics, digestive system development, intestinal microflora and some blood parameters of broiler chicks. Europ Poult Conf. Verona, Italy. 2006.

[80] Feizi A, Bijanzad P, Kaboli K. Effects of thyme volatile oils on performance of broiler chickens. Eur J Exp Biol 2013; 3: 250-4.

[81] Ayoola MA, Adedeji OA, Oladepo AD. Effects of dietary thyme leaf on broiler growth performances, carcass characteristics and cooking yield of the meat. GJSR Journal 2014; 2: 47-50.

[82] Blair R. Nutrition and Feeding of Organic Poultry. CAB International, Wallingford, UK, 314 pp, Animal Feed Science and Technology 2008; 01/2009; 151: 172-173.

[83] Rafiee A, Rahimian Y, Zamani F, Asgarian F. Effect of use ginger (*Zingiber officinale*) and thymus (*Thymus vulgaris*) extract on performance and some hematological parameters on broiler chicks. Sci Agric 2013; 4: 20-5.

[84] Alfaig E, Angelovicova M, Kral M. Effect of probiotics and thyme essential oil on carcass parameters of broiler chickens. Scientific Papers: Anim Sci and Bio 2013; 46: 50-2.

[85] Legarreta IG, Hui YH. Poultry Science and Technology, vol. 1 In: Primary Processing, Published by John Wiley and Sons. Inc 2010; pp. 101.

[86] Duval ELB, Millet N, Remignon H. Broiler meat quality: effect of selection for increased carcass quality and estimates of genetic 1999.

[87] Shabaan M. Effect of using thyme (*Thymus vulgaris* L) and cumin (*Cuminum cyminum* L) seeds for improving the utilization of low energy broiler diet. Egypt Poult Sci J 2012; 32: 579-91.

[88] Zargari A. Medical plants. 2nd ed. Tehran University Press 2001; pp. 25-36.

[89] Varel VH. Carvacrol and thymol reduce swine waste odor and pathogens: stability of oils. Curr Microbiol 2002; 44(1): 38-43.
[http://dx.doi.org/10.1007/s00284-001-0071-z] [PMID: 11727039]

[90] Placha I, Takacova J, Ryzner M, *et al.* Effect of thyme essential oil and selenium on intestine integrity and antioxidant status of broilers. Br Poult Sci 2014; 55(1): 105-14.
[http://dx.doi.org/10.1080/00071668.2013.873772] [PMID: 24397472]

[91] Feizi A, Nazeri M. The effect of thyme essential oils (*Thymus vulgaris*) in the vaccination reactions on broiler chicks. Adv Environ Biol 2011; 5: 1912-5.

[92] Zadeh T, Rahimi SH, Karimi Torshizi MA, Omidbaigi R. The effects of *Thymus vulgaris* L., *Echina purpuria* (L.) Moench. *Allium sativum* L. extracts and virginiamycin antibiotic on intestinal microflora population and immune system in broilers. Iranian Journal of Medicinal and Aromatic Plants. Tahqiqat-i Giyahan-i Daruyi va Muattar-i Iran 2009; 25: 39-48.

CHAPTER 5

Rosmarinus officinalis (Rosemary)

Mohamed E. Abd El-Hack[1,*], **Mahmoud Alagawany**[1], **Youssef A. Attia**[2,3,*], **Adel D. Al-qurashi**[3], **Abdulmohsen H. Alqhtani**[4], **Hossam A. Shahba**[5], **Asmaa F. Khafaga**[6], **Vincenzo Tufarelli**[7], **Maria Cristina de Oliveira**[8] and **Omer H.M. Ibrahim**[3,9]

[1] *Poultry Department, Faculty of Agriculture, Zagazig University, Zagazig-44519, Egypt*

[2] *Animal and Poultry Production Department, Faculty of Agriculture, Damanhour University, Damanhour-22713, Egypt*

[3] *Sustainable Agriculture Production Research Group, Agriculture Department, Faculty of Environmental Sciences, King Abdulaziz University, Jeddah-21589, Saudi Arabia*

[4] *Department of Animal Production, College of Food and Agricultural Sciences, King Saud University, Riyadh, Saudi Arabia*

[5] *Rabbit, Turkey and Water Fowl Research Department, Animal Production Research Institute, Agriculture Research Center, Dokki, Giza-3751310, Egypt*

[6] *Department of Pathology, Faculty of Veterinary Medicine, Alexandria University, Apis, Alexandria, 21944, Egypt*

[7] *Department of Precision and Regenerative Medicine and Jonian Area (DiMePRe-J), Section of Veterinary Science and Animal Production, University of Bari 'Aldo Moro' s.p. Casamassima km 3, 70010-Valenzano, Italy*

[8] *University of Rio Verde, Faculty of Veterinary Medicine, Rio Verde, GO, Brazil*

[9] *Department of Floriculture, Faculty of Agriculture, Assiut University, Egypt*

Abstract: Rosemary is a phytogenic aromatic plant, abundant in phenolic acids, such as caffeic (0.69-1.0 mg/g) and rosmarinic acids (16.77-29.91 mg/g), flavonoids, and diterpenes. Additional components of rosemary include camphor, 1,8-cineole, and α-pinene. Rosmarinic acid (RA) exhibits antioxidant, antiviral, antibacterial, anti-inflammatory, and antimutagenic properties. Furthermore, rosemary significantly reduced the peroxidation of unsaturated lipids and enhanced the levels of reduced glutathione and antioxidant enzyme activities in the kidney and testis compared to aspartame controls. Additionally, rosemary essential oil augments the resistance of rat

---

* **Corresponding authors Mohamed E. Abd El-Hack and Youssef A. Attia:** Poultry Department, Faculty of Agriculture, Zagazig University, Zagazig-44519, Egypt and Sustainable Agriculture Production Research Group, Agriculture Department, Faculty of Environmental Sciences, King Abdulaziz University, Jeddah-21589, Saudi Arabia and Department of Animal Production, College of Food and Agricultural Sciences, King Saud University, Riyadh, Saudi Arabia;
E-mails: dr.mohamed.e.abdalhaq@gmail.com, yaattia@kau.edu.sa

Youssef A. Attia, Mohamed E. Abd El-Hack, Mahmoud M. Alagawany & Asmaa Sh. Elnaggar
(Eds.)

hepatocytes against DNA-damaging oxidative agents and serves as an effective free radical scavenger. Caffeic and rosmarinic acids have demonstrated potential in the treatment of inflammatory diseases and hepatotoxicity. Rosemary is also rich in phytochemical derivatives such as triterpenes, flavonoids, and polyphenols. This review will focus on the beneficial effects of rosemary and its application in improving the productive performance and health of livestock.

Keywords: Bioactive substances, Immunity, Livestock, Productive performance, Reproductive, Rosemary.

INTRODUCTION

Rosmarinus officinalis L., commonly known as rosemary, from the family *Lamiaceae* [1] is a perennial shrub with fragrant leaves (Fig. **1**). It is popular for culinary use because its leaves have a pungent and slightly bitter taste [2].

Fig. (1). Rosemary plant.
Source: Contribution of Hans by Pixabay [3].

Rosemary is a phytogenic aromatic plant, rich in phenolic acids, such as caffeic (0.69-1.0 mg/g) and rosmarinic acids (16.77-29.91 mg/g), flavonoids, and diterpenes [4, 5]. Other components of rosemary include camphor, 1,8-cineole, and α-pinene [6].

Rosmarinic acid is an ester of 3,4-hi hydroxyphenyl lactic acid and caffeic acid [7], which has antioxidant [4, 8, 9], antiviral [10 - 12], antibacterial [13, 14], anti-inflammatory [4, 15, 16], and antimutagenic [17] properties.

Moreover, rosemary significantly decreased the elevation in peroxidation of unsaturated lipids and boosted the levels of reduced glutathione and antioxidant enzyme actions in the kidney and testis compared to aspartame controls [4, 18, 19]. Furthermore, Isles *et al.* [20] and Harvàthová *et al.* [21] stated that rosemary essential oil enhances the resistance of rat hepatocytes to DNA-damaging Caffeic and rosmarinic acids, which have the potential to treat inflammatory diseases and hepatotoxicity [22, 23]. Rosemary is also rich in phytochemical derivatives such as triterpenes, flavonoids, and polyphenols [4, 24, 25] and can protect against oxidative stress as an effective agent for scavenging free radicals [26, 27]. Fareed *et al.* [28] showed that rosemary essential oil restored normal kidney function in diabetic rats, reducing the blood glucose, urea, creatinine, uric acid, malondialdehyde, and catalase concentrations to normal levels.

Rosemary contains many bioactive compounds such as phenolic acids; polyphenols, phenolic diterpenoid bitter substances, tripenoid acids, flavonoids; 1.2 to 2.5% volatile oil, and tannins [29, 30]. Epirosmanol phenolic diterpenes of rosemary prevent peroxidation of unsaturated lipids [31]. Rosemary also contains essential components such as carnosic acid, carnosal, and caffeic acid, and its derivatives such as rosmarinic acid. These constituents displayed potent antioxidant activities. Moreover, rosemary and its components have therapeutic potential for peptic ulcers, bronchial asthma, liver toxicity, prostate disorders, stroke, inflammatory diseases, leukemia, atherosclerosis, ischemic heart disease, and cataracts. It is widely utilized in food processing as a flavoring agent and spice [22]. The main bioactive compounds of rosemary are illustrated in Fig. (**2**).

Effects on Livestock Productive Performance

The beneficial effects of rosemary supplementation on the productive performance of animals could be ascribed to its bioactive substances, such as borneol, carnosol, carnosic acid, and caffeic acid. These substances stimulate the digestion of nutrients by increasing digestive enzymes and boosting nutrient utilization through heightened liver function [32, 33].

In addition, 1% rosemary can be used as an antimicrobial agent for broiler chickens [34], decreasing the presence of *E. coli, Salmonella typhimurium, Bacillus cereus*, and *Staphylococcus aureus* [35] and enhancing animal health and performance [36]. The antibacterial effect is due to the interaction with the cell membrane, which changes the genetic material and the transport of electrons, allowing the leakage of cellular components and the loss of membrane functionality and structure [37].

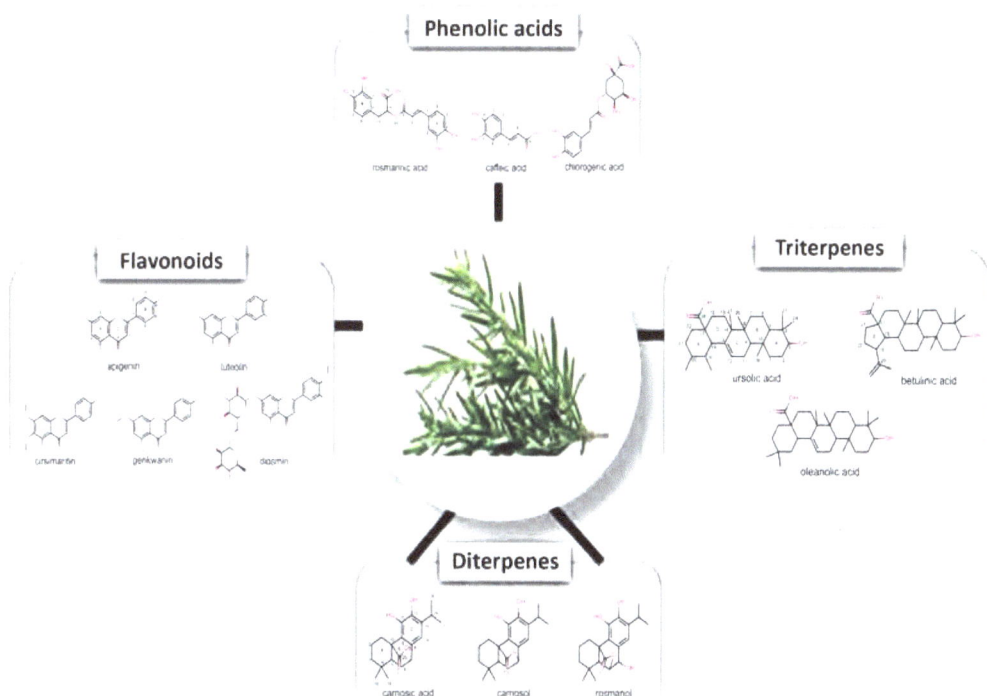

Fig. (2). Main bioactive compounds of rosemary.

Rosemarinus officinalis extract (ROE) is a well-known phytogenic agent for farm animals and has been studied as a feed additive in poultry [4, 38 - 40]. Supplementation of broiler diets with 0.5% rosemary leaves increased the growth rate compared to that in the control group [41]. Moreover, Petricevic *et al.* [42], Hosseinzadeh *et al.* [43], Nameghi *et al.* [44], and Yao *et al.* [45] showed that various rosemary concentrations significantly improve the growth performance of broilers' and ducks.' Comparable results were observed by Manafi *et al.* [46], who revealed that the body weight gain of broilers at 42 d of age increased in the group supplemented with 500 mg ROE alone compared to the other groups (control: aflatoxins B1(AF) and AF+ rosemary extract).

In studies on laying hens, Bölükbasi *et al.* [47] observed that rosemary oil supplementation (200 mg/kg diet) reduced the feed intake and feed: gain ratio and increased the egg weight of hens. The same trend was noted by Garcia *et al.* [48], with a better laying rate and egg mass due to the inclusion of a 200 mg/kg diet of rosemary oil. However, the authors did not report any improvements in feed intake, feed conversion rate, or egg quality. In addition, Kedir *et al.* [49] reported that rosemary leaf meal supplementation at 3.5% and 5.2% improved laying performance and quality of fresh and stored eggs.

Supplementation with 100, 200, or 400 mg/kg of ROE in pig diets increased body weight and daily gain, and a linear increase in feed intake was observed. Digestibility of crude protein and energy were also linearly increased, as villus height, and villus: crypt ratio in the jejunum and ileum [50].

Elazab *et al.* [51] reported an improvement in growth performance and feed utilization in rabbits fed diets containing 0.5% rosemary essential oil. Attia *et al.* [52] reported that the inclusion of rosemary leaves powder (5 and 10 g/kg) in diets for growing rabbits from 28 to 91 days of age improved the feed conversion ratio, compared to the control, with the best values obtained with 5 g/kg of rosemary leaves. Mousa *et al.* [53] found that the use of chamomile + rosemary (5 ml/L water) increased the growth rate and feed conversion rate as compared to the use of chamomile (5ml/L water) or rosemary (5 and 10 ml/L water) groups.

Naiel *et al.* [54] evaluated the supplementation of 0.1%, 0.25%, and 0.5% ROE in diets containing 2.5 mg/kg of aflatoxin B1 for fish. The authors demonstrated that 0.5% ROE improved the growth performance of the animals to the same level as that of the control group. More recently, Naiel *et al.* [55] also noted better growth performance of Nile tilapia fingerlings by adding 10 g/kg of rosemary leaf powder.

The addition of 28 g/d of ROE in the diets of dairy cows resulted in higher milk production (40.94 *vs.* 41.76 kg/d) and milk protein yields (1.37 *vs.* 1.42 kg/d) than the control group [56]. Souza *et al.* [57] verified the adverse effects of ROE supplementation on Nellore heifers. The animals had lower feed intake, daily gain, and feed efficiency due to the inclusion of 4 g/animal/day of ROE. According to Yagoubi *et al.* [58], 200 g rosemary residual and 300 g straw provided to ewes for 90 days improved the intake and digestibility of crude protein in the diets. Farghaly and Abdullah [59] also demonstrated that the dietary supplementation of 2.5% rosemary essential oil can improve the nutrient digestibility in lambs. Tawfeeq *et al.* [60], however, did not report any beneficial effect on ruminant *in-vitro* digestibility due to the supplementation with rosemary leaves.

Nonetheless, rosemary supplementation did not significantly affect the growth performance of broiler chickens [50], pigs [61], sea bass [62], rats [63, 64], rabbits [65, 66], goats [67], or laying hens [68]. The advantages of rosemary fortification in poultry are summarized in Fig. (**3**).

- Reduce total lipids, cholesterol, triglyceride and total cholesterol
- Scavenging of free radical-induced tissue injury

- Improve growth performance and body weight gain by stimulate the antimicrobial actions and digestion by increasing digestive enzymes,
- Boosting nutrient utilizations through heightened liver functions

- Increased feed intake due to the appetizing effect of bioactive substances

- Positively influenced immunity by increase total proteins, albumin, and globulin and activate immune functions OS WBCs, lymphocyte proliferation, phagocytosis, red blood cell and hemoglobin and preventing enzymatic oxidation

- Improve villus height and crypt depth
- Reduce goblet cells density in small intestine
- Enhance the intestinal microbiota population, intestinal morphology, immune response, and plasma biochemistry parameters in broiler chickens

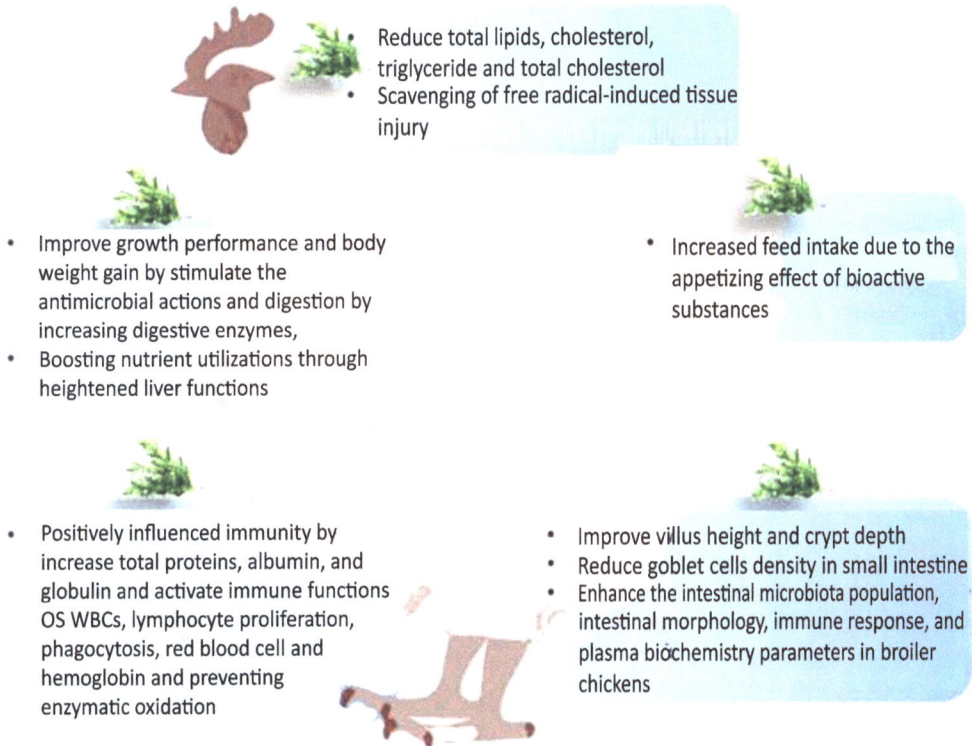

Fig. (3). Advantages of dietary rosemary supplementation in poultry.

Effects on Carcass Characteristics and Meat Quality

The beneficial effects of rosemary on feed intake and weight gain are generally reflected in carcass traits, as shown by Tollba [36], who reported higher values of carcass dressing and giblets in 6-week-old broilers supplemented with rosemary.

However, Al-Shuwaili [69] reported no changes in carcass yield of chickens fed different concentrations of rosemary, thyme, or a mixture of rosemary and thyme compared to the control group. Attia [70] also reported no effect on carcass traits in broilers due to different rosemary leaf supplementation levels (0.25, 0.50, 0.75, and 1.0%), except for the abdominal fat percentage, which was lower due to the supplementation of 0.5% rosemary leaves, compared with the control. Similar results were obtained by Attia *et al.* [70], who found that rosemary leaf at 5 and 10 g/kg diet did not affect carcass traits, abdominal fat, and body organs proportional to growing rabbits and by El-Gogary *et al.* [65] who verified no influence of rosemary essential oil (0.25, 0.50, and 0.75 g/kg diet) on the carcass yield of broilers. However, Yagoubi *et al.* [58] used rosemary residues in the diet of lambs and did not observe any differences in the weight or yield of the hot and cold carcasses.

Several plants have been used to improve animal product quality due to the need to meet consumers' higher demand for safe and high-quality food [58]. The improvement depends on the content of bioactive compounds, such as components with antioxidant properties, to reduce meat oxidation and increase the meat shelf life.

Lopez-Bote *et al.* [71] demonstrated that the malondialdehyde (MDA) concentrations in meat from chickens fed diets supplemented with rosemary and sage extracts ranged from 0.30 to 0.35 mg MDA/kg meat and were significantly reduced in comparison to those in chickens fed the control unenriched diet.

The effects of rosemary leaves and essential oil in turkey diets were studied by Jridi *et al.* [72], who found that the addition of essential oil did not affect the sausage texture and color of meat. However, increased taste and aroma scores reduced total bacterial counts and TBARS values during chill storage, compared with the control. Ortuño *et al.* [73] evaluated the effect of 200 or 400 mg of rosemary extract in lambs' diets on their meat quality. The authors reported that rosemary extracts delayed lean and fat discoloration, lipid oxidation, odor deterioration, and microbial spoilage, thereby increasing the shelf life of the meat. Delayed lipid oxidation in pork meat sprayed with rosemary extract has been described by Perlo *et al.* [74].

Better nutritional value for the meat of animals receiving rosemary products in their diets has been described. Elazab *et al.* [51] observed that rabbit meat showed lower levels of cholesterol when the animals received dietary rosemary essential oil (0.5%) compared to the control group (48.20 *vs.* 50.51 mg/100 g *longissimus dorsi* muscle and 64.39 *vs.* 66.06 mg/100 g of hind leg muscle, respectively). Naiel *et al.* [54, 55] reported higher levels of protein and minerals in the meat of Nile tilapia receiving rosemary. In addition, Liotta *et al.* [75] reported that rosemary extract (1 g/kg) increased the polyunsaturated fatty acid content of pork meat, which is considered a functional additive in animal feeding.

Blood Parameters

Rosmarinus officinalis has a hypoglycemic effect due to the regeneration of pancreatic β-cells and increased blood insulin levels and glucose uptake [76 - 78]. The reduction in lipid levels may be due to lower absorption of dietary fat and higher fecal fat excretion [79]. In addition, rosemary inhibits hormone-sensitive lipase and pancreatic lipase, thereby reducing triglyceride hydrolysis [80].

Carnosol, a component of *Rosmarinus officinalis,* inhibits diacylglycerol acyltransferase, an enzyme that catalyzes the formation of triglycerides [77]. Another possibility is that rosemary increases LDL receptor expression and uptake of LDL by hepatocytes [76], and that the decline in blood sugar and increase in insulin secretion regulate and inhibit lipolysis in adipocytes [81].

Al-Jamal and Alqadi [82] demonstrated that diabetic rats receiving rosemary for four weeks showed a decrease of 20% in glucose level, 22% in total cholesterol, 24% in triglycerides, and 27% in LDL, and an increase of 18% in HDL. Al-Sheyab *et al.* [83] found that the oral supplementation of extract of rosemary at 100 mg/kg per d for 15 d to high cholesterol-fed mice caused significant decreases in plasma total cholesterol (- 68.57%), LDL-C (- 56.34%), and triglycerides (- 182.61%) in comparison to high cholesterol-fed mice, while, there was an elevation (38.53%) in HDL as compared to high cholesterol mice.

Diets containing 1 and 3 g/kg rosemary extract for rainbow trout resulted in lower blood levels of ALT, AST, glucose, total cholesterol, and triglycerides [32]. Mousa *et al.* [53], supplementing rosemary extract at 10 ml/L to the drinking water of growing rabbits, observed an increase in the hemoglobin and hematocrit values and the number of white blood cells as compared with the control group (7.07 *vs.* 3.65×10^3, respectively). In addition, Elazab *et al.* [51] studied rabbits supplemented with rosemary essential oil (0.5% diet); however, there were no differences among the values for LDL and HDL.

Ghazalah and Ali [41] studied the effect of rosemary on broilers' diets and found that blood levels of glucose, total lipids, total cholesterol, LDL, creatinine, and uric acid were reduced as compared to the control. Lower blood levels of total cholesterol and triglycerides were found by Nameghi *et al.* [44] in broilers fed diets containing rosemary powder (7.5 g/kg diet), which resulted in lower levels of triglycerides, total cholesterol, and LDL fraction. The values of HDL, AST, and ALT did not differ from those of the control group.

Hosseinzadeh *et al.* [43] reported that serum total protein, albumin, glucose, total cholesterol, and triglyceride concentrations were reduced in groups supplemented with 5 and 10 ml/L of rosemary extract compared to the control group in broilers.

Owing to its antioxidant effects, rosemary acts as a hepatoprotective [84] and nephroprotective [85, 86] agent, preserving cellular integrity and increasing the activity of antioxidant enzymes [87]. Free radical damage, evaluated by MDA levels as the final product of lipid peroxidation, may cause the depletion of antioxidant reserves.

Diets supplemented with 1 and 3 g/kg rosemary extract caused an increase in the antioxidant enzymes catalase and superoxide dismutase [49]. The same effect was described by Rasoolijazi *et al.* [88], who found an increase in the activities of glutathione peroxidase, superoxide dismutase, and catalase in the hippocampus of rats treated with 100 mg/kg rosemary extract, and by Afonso *et al.* [64] in hypercholesterolemic rats receiving an aqueous extract of rosemary (140 mg/kg). The authors mentioned that the extract increased the activities of antioxidant enzymes and reduced the amount of thiobarbituric acid-reactive substances.

El-Gindy *et al.* [66] reported that the inclusion of rosemary essential oil in the diets of growing rabbits resulted in increased red blood cell counts, hemoglobin content, higher activities of the antioxidant enzymes superoxide dismutase, glutathione, and catalase, and increased total antioxidant capacity. The same effect was observed by Liu *et al.* [89] in broilers fed diets containing 250, 500, 750, and 1000 mg/kg rosemary extract. However, different results have been reported by several researchers. Nameghi *et al.* [44] did not find differences in the activities of the antioxidant enzymes in broilers receiving rosemary powder (7.5 g/kg diet).

Effects on Immunity Profile

Phytogenics may improve immunity by enhancing the intestinal microbiota and its antioxidant components [90], which may increase serum immunoglobulin levels [91], neutrophil adhesion, number of T-helper and T-suppressor cells, and activation of macrophages [92]. According to Hashemipour *et al.* [93] and Alagawany and Abd El-Hack [68], medicinal plants and their components can activate immune functions, such as white and red blood cell production, lymphocyte proliferation, phagocytosis, and hemoglobin synthesis.

Lysozyme, a bactericidal enzyme, and the alternative complement pathway (ACH50) are indicators of an immediate innate immune system that promotes pathogen resistance [55]. Higher values for lysozyme activity, total immunoglobulin, and white blood cells were noted by Karatas *et al.* [32] in rainbow trout supplemented with a 3 g/kg diet than in the control group. Naiel *et al.* [55] cited that rosemary leaf powder at 5 and 10 mg/kg diet caused an increase in lysozyme and ACH50 in Nile tilapia, indicating an immunomodulatory effect in the fish.

Better production of interleukine-2, IgM, IgG, and IgA may occur in broilers and ducks fed diets containing rosemary extract (250, 500, 750, and 1000 mg/kg) [45, 89]. Alagawany and Abd El-Hack [68] found that adding rosemary to layer diets might improve serum IgM concentrations compared with the control diet. According to Hosseinzadeh *et al.* [43], rosemary essential oils enhance antibody titers against sheep red blood cells and Newcastle disease virus and increase the total plasma protein and globulin levels of broilers.

Shokrollahi *et al.* [67] found that globulin and white blood cell levels were increased in goats' kids receiving 200 or 400 mg/kg body weight compared to the control. Leukocytes were the highest in animals receiving rosemary extract at 200 and 300 mg/kg body weight compared to the control group. However, no alterations were observed in the neutrophils or lymphocytes among the various groups. The authors concluded that rosemary extract positively influenced immunity owing to its ability to develop and defend cells and prevent non-enzymatic oxidation.

The results on the effects of rosemary on immunity are contradictory, as shown by Ghazalah *et al.* [41]; however, they did not find differences in antibody titers against sheep red blood cells in broilers receiving diets supplemented with 0.5, 1, and 2% rosemary leaf powder. Soltani *et al.* [94] and Yang *et al.* [50] also did not report an effect on antibody titers against Newcastle disease, Influenza disease, and Infectious Bursal disease in broilers due to the use of rosemary powder or rosemary ethanolic extract. El-Gindy *et al.* [66] also reported a lack of influence of rosemary essential oil on rabbit immunity.

Effects on Reproductive Performance of Livestock

Owing to the antioxidant property of *Rosmarinus officinalis,* its administration to animals prevents free radical damage to the cellular membrane and raises normal cell viability [95]. However, the effects of rosemary on fertility remain controversial because they can reduce gonadal damage induced by free radicals at a low dose; however, high-dose rosemary may have adverse effects [96].

Impairments in the weight of reproductive organs, motility and density of sperm, several germinal and interstitial cell types, and serum levels of testosterone, FSH, and LH were found by Nusier *et al.* [97] in adult male rats fed diets with 500 mg/kg body weight of ethanolic rosemary extract. These effects were not observed when the rats received 250 mg/kg of the extract, showing a dose-dependent effect.

Improvements in testosterone, FSH, and LH levels have been reported in male rats by Hamzah *et al.* [98] and in male rabbits by Hasan and Al-Rikaby [99]. However, Heidari-Vala *et al.* [100] suggested that the rosemary extract has antiandrogenic activity at doses of 50 and 100 mg/kg of body weight. However, these doses do not affect spermatogenesis.

Effect on Milk Production

Milk production in ruminants may be increased by rosemary extracts because their phenolic compounds reduce starch and protein degradation in the rumen, representing a nutritive apport for use in milk production [101, 102]. In addition, propionic acid production is increased by rosemary, which is used in gluconeogenesis [103] and provides energy to support milk production.

The use of rosemary extracts at 1200 mg/animal/day increased milk production and protein and lactose yields in ewes. The use of 600 mg/animal/day of the extract did not improve the results compared to the control animals [101]. Rosemary leaves or rosemary essential oil increases the daily milk production and the level of protein in goats' milk, improving the kid's body weight. These effects are attributed to phenolic compounds, which may enhance ruminal fermentation [4, 102]. Including rosemary in goat diets improves propionic acid production, milk production, milk component yield, and fatty acid profile in milk [103].

CONCLUSION AND REMARKS

Several botanicals, including rosemary, can potentially improve the health and productive performance of livestock because of their beneficial effects, such as increasing digestive secretion, improving intestinal microbiota, boosting the immune system, and improving the shelf life and nutritional value of meat products. Rosemary, along with other botanicals, has garnered significant attention for livestock management because of its multifaceted benefits. Additionally, the positive impact of rosemary on the intestinal microbiota helps maintain a balanced gut environment, which is crucial for optimal digestion and nutrient utilization. This botanical also exhibits immunomodulatory properties, strengthening the animals' natural defense mechanisms against pathogens and environmental stressors.

Beyond its effects on live animals, rosemary continues to provide post-harvest benefits. Its potent antioxidant properties extend the shelf life of meat products by inhibiting lipid oxidation and microbial growth. This not only ensures food safety, but also preserves the nutritional quality of the meat, maintaining its flavor, color, and texture for longer periods. Furthermore, the incorporation of rosemary in animal feed can potentially enhance the nutritional profile of the resulting meat products, offering consumer meat with improved fatty acid composition and increased antioxidant content. These combined effects make rosemary a promising natural additive for livestock production, addressing both animal welfare and food quality concerns.

IMPLICATIONS

Rosemary, a common herb, has been studied extensively for its potential benefits in animal farming. Research has shown that rosemary extracts and oils may help improve the health, productivity, and quality of livestock products. These benefits are due to the ability of rosemary to act as an antioxidant, fight microbes, and boost the immune response. The antioxidant, antimicrobial, and immunomodulatory properties of rosemary extracts and essential oils may lead to their increased use as natural feed additives in animal agriculture. This may help reduce reliance on synthetic additives and antibiotics, addressing consumer demands for more natural production methods. However, the varying and sometimes contradictory results across different studies indicate that more studies are needed to determine the optimal dosages, forms of administration, and specific applications for different livestock species. Additionally, the potential anti-androgenic effects at high doses suggest that caution is needed when using rosemary supplements, particularly in breeding animals. Overall, rosemary shows promise as a multifunctional feed additive; however, its use requires careful consideration of dosage and target outcomes to maximize benefits while avoiding potential negative effects. In summary, although rosemary shows promise as a versatile feed additive in animal farming, its use requires careful planning to maximize benefits and avoid potential negative effects.

REFERENCES

[1]   Dias ISSP, Menezes Filho ACP, Porfiro CA. O uso do óleo essencial de *Rosmarinus officinalis* L. no paciente com Alzheimer. Braz J Sci 2022; 1: 66-96.
[http://dx.doi.org/10.14295/bjs.v1i3.117]

[2]   Britannica. Rosemary. 2023. (accessed on August 4, 2023). Available from: https://www.britannica.com/plant/rosemary

[3]   Pixabay. Rosemary. Contribution of Hans. 2015. Available from: https://pixabay.com/pt/photos/alecrim-flores-azul-tolet-1090419/

[4]   Attia YA. Recent approaches to replace antibiotics in animal nutrition as alternative growth promoters Novel Techniques in Nutrition & Food Science 2018; 1 (5). NTNF.000523.2018. Available from: http://crimsonpublishers.com/
[http://dx.doi.org/10.31031/NTNF.2019.01.000523]

[5]   Kheiria H, Mounir A, María Q, José JM, Bouzid S. Total phenolic content and polyphenolic profile of Tunisian rosemary (*Rosmarinus officinalis* L.) residues. El-Shemy, HA Natural drugs from plants. IntechOpen 2021.
[http://dx.doi.org/10.5772/intechopen.101762]

[6]   Sienkiewicz M, Łysakowska M, Pastuszka M, Bienias W, Kowalczyk E. The potential of use basil and rosemary essential oils as effective antibacterial agents. Molecules 2013; 18(8): 9334-51.
[http://dx.doi.org/10.3390/molecules18089334] [PMID: 23921795]

[7]   Elufioye TO, Habtemariam S. Hepatoprotective effects of rosmarinic acid: Insight into its mechanisms of action. Biomed Pharmacother 2019; 112: 108600.
[http://dx.doi.org/10.1016/j.biopha.2019.108600] [PMID: 30780110]

[8]   Adomako-Bonsu AG, Chan SLF, Pratten M, Fry JR. Antioxidant activity of rosmarinic acid and its principal metabolites in chemical and cellular systems: Importance of physico-chemical characteristics. Toxicol. *In vitro* 2017; 40: 248-55.
[http://dx.doi.org/10.1016/j.tiv.2017.01.016] [PMID: 28122265]

[9]   Aldoghachi FEH, Noor Al-Mousawi UM, Shari FH. Antioxidant activity of rosmarinic acid extracted and purified from *Mentha piperita.* Arch Razi Inst 2021; 76(5): 1279-87.
[http://dx.doi.org/10.22092/ari.2021.356072.1770] [PMID: 35355734]

[10]   Lin WY, Yu YJ, Jinn TR. Evaluation of the virucidal effects of rosmarinic acid against enterovirus 71 infection *via in vitro* and *in vivo* study. Virol J 2019; 16(1): 94.
[http://dx.doi.org/10.1186/s12985-019-1203-z] [PMID: 31366366]

[11]   Jheng JR, Hsieh CF, Chang YH, *et al.* Rosmarinic acid interferes with influenza virus A entry and replication by decreasing GSK3β and phosphorylated AKT expression levels. J. Microbiol 2022; Immunol. Infect. 55: 598-610.
[http://dx.doi.org/10.1016/j.jmii.2022.04.012]

[12]   Panchal R, Ghosh S, Mehla R, *et al.* Antiviral activity of rosmarinic acid against four serotypes of dengue virus. Curr Microbiol 2022; 79(7): 203.
[http://dx.doi.org/10.1007/s00284-022-02889-3] [PMID: 35612625]

[13]   Zhang J, Cui X, Zhang M, Bai B, Yang Y, Fan S. The antibacterial mechanism of perilla rosmarinic acid. Biotechnol Appl Biochem 2022; 69(4): 1757-64.
[http://dx.doi.org/10.1002/bab.2248] [PMID: 34490944]

[14]   Kernou ON, Azzouz Z, Madani K, Rijo P. Application of rosmarinic acid with its derivatives in the treatment of microbial pathogens. Molecules 2023; 28(10): 4243.
[http://dx.doi.org/10.3390/molecules28104243] [PMID: 37241981]

[15]   Luo C, Zou L, Sun H, *et al.* A review of the anti-inflammatory effects of rosmarinic acid on inflammation diseases. Front Pharmacol 2020; 11: 153.
[http://dx.doi.org/10.3389/fphar.2020.00153] [PMID: 32184728]

[16]   Marinho S, Illanes M, Ávila-Román J, Motilva V, Talero E. Anti-inflammatory effects of rosmarinic acid-loaded navovesicles in acute colitis through modulation of NLRP3 inflammasome. Biomolecules 2021; 11(2): 162.
[http://dx.doi.org/10.3390/biom11020162] [PMID: 33530569]

[17]   Kakouri E, Nikola O, Kanakis C, *et al.* Cytotoxic effect of *Rosmarinus officinalis* extract on glioblastoma and rhabdomyosarcoma cell lines. Molecules 2022; 27(19): 6348.
[http://dx.doi.org/10.3390/molecules27196348] [PMID: 36234882]

[18]   Pérez-Fons L, Aranda FJ, Guillén J, Villalaín J, Micol V. Rosemary (*Rosmarinus officinalis*) diterpenes affect lipid polymorphism and fluidity in phospholipid membranes. Arch Biochem Biophys 2006; 453(2): 224-36.
[http://dx.doi.org/10.1016/j.abb.2006.07.004] [PMID: 16949545]

[19]   Hozayen WG, Soliman HAE, Desouky EM. Potential protective effects of rosemary extract, against aspartame toxicity in male rats. J Int Acad Res Multidisc 2014; 2: 111-25.

[20]   Isles RC, Choy NLL, Steer M, Nitz JC. Normal values of balance tests in women aged 20-80. J Am Geriatr Soc 2004; 52(8): 1367-72.
[http://dx.doi.org/10.1111/j.1532-5415.2004.52370.x] [PMID: 15271128]

[21]   Horváthová E, Slameňová D, Navarová J. Administration of rosemary essential oil enhances resistance of rat hepatocytes against DNA-damaging oxidative agents. Food Chem 2010; 123(1): 151-6.
[http://dx.doi.org/10.1016/j.foodchem.2010.04.022]

[22]   Katerinopoulos H, Pagona G, Afratis A, Stratigakis N, Roditakis N. Composition and insect attracting activity of the essential oil of *Rosmarinus officinalis.* J Chem Ecol 2005; 31(1): 111-22.
[http://dx.doi.org/10.1007/s10886-005-0978-0] [PMID: 15839484]

[23]   Khafaga AF, Abd El-Hack ME, Taha AE, Elnesr SS, Alagawany M. The potential modulatory role of herbal additives against Cd toxicity in human, animal, and poultry: a review. Environ Sci Pollut Res Int 2019; 26(5): 4588-604.
[http://dx.doi.org/10.1007/s11356-018-4037-0] [PMID: 30612355]

[24]   Abo Ghanima MM, Elsadek MF, Taha AE, *et al.* Effect of housing system and rosemary and cinnamon essential oils on layers performance, egg quality, haematological traits, blood chemistry, immunity, and antioxidant. Animals (Basel) 2020; 10(2): 245.
[http://dx.doi.org/10.3390/ani10020245] [PMID: 32033082]

[25]   Seidavi A, Tavakoli M, Asroosh F, *et al.* Antioxidant and antimicrobial activities of phytonutrients as antibiotic substitutes in poultry feed. Environ Sci Pollut Res Int 2022; 29: 5006-31.
[PMID: 34811612]

[26]   Alavi MS, Fanoudi S, Ghasemzadeh Rahbardar M, Mehri S, Hosseinzadeh H. An updated review of protective effects of rosemary and its active constituents against natural and chemical toxicities. Phytother Res 2021; 35(3): 1313-28.
[http://dx.doi.org/10.1002/ptr.6894] [PMID: 33044022]

[27]   Mohamed ME, Younis NS, El-Beltagi HS, Mohafez OM. The synergistic hepatoprotective activity of rosemary essential oil and curcumin: the role of the MEK/ERK pathway. Molecules 2022; 27(24): 8910.
[http://dx.doi.org/10.3390/molecules27248910] [PMID: 36558044]

[28]   Fareed SA, Yousef EM, Abd El-Moneam SM. Assessment of effects of rosemary essential oils on the kidney pathology of diabetic adult male albino rats. Cureus 2023; 15(3): e35736.
[http://dx.doi.org/10.7759/cureus.35736] [PMID: 37016650]

[29]   Leung AY, Foster S. Encyclopedia of common natural ingredients used in food, drugs, and cosmetics. 1996.

[30]   Gharejanloo M, Mehri M, Shirmohammad F. Effect of different levels of turmeric and rosemary essential oils on performance and oxidative stability of broiler meat. Iran J Appl Anim Sci 2017; 7(4): 655-62.

[31]   Zheng W, Wang SY. Antioxidant activity and phenolic compounds in selected herbs. J Agric Food Chem 2001; 49(11): 5165-70.
[http://dx.doi.org/10.1021/jf010697n] [PMID: 11714298]

[32]   Karataş T, Korkmaz F, Karataş A, Yildirim S. Effects of rosemary (*Rosmarinus officinalis*) extract on growth, blood biochemistry, immunity, antioxidant, digestive enzymes and liver histopathology of rainbow trout, *Oncorhynchus mykiss.* Aquacult Nutr 2020; 26(5): 1533-41.
[http://dx.doi.org/10.1111/anu.13100]

[33]   Hernández F, Madrid J, García V, Orengo J, Megías MD. Influence of two plant extracts on broilers performance, digestibility, and digestive organ size. Poult Sci 2004; 83(2): 169-74.
[http://dx.doi.org/10.1093/ps/83.2.169] [PMID: 14979566]

[34]   Mohammed A GA, Mohammed MF, Hamood MF, Jameel YJ. The effect of anise and rosemary on the microbial balance in gastro intestinal tract for broiler chicks. Int J Poult Sci 2008; 7(6): 610-2.
[http://dx.doi.org/10.3923/ijps.2008.610.612]

[35]   Burt S. Essential oils: their antibacterial properties and potential applications in foods—a review. Int J Food Microbiol 2004; 94(3): 223-53.
[http://dx.doi.org/10.1016/j.ijfoodmicro.2004.03.022] [PMID: 15246235]

[36]   Tollba A. Reduction of broilers intestinal pathogenic micro-flora under normal or stressed condition. Egypt Poult Sci 2010; 30: 249-70.

[37]   Manilal A, Sabu KR, Woldemariam M, *et al.* Antibacterial activity of *Rosmarinus officinalis* against multidrug-resistant clinical isolates and meat-borne pathogens. Evid Based Complement Alternat Med 2021; 2021: 1-10.

[http://dx.doi.org/10.1155/2021/6677420] [PMID: 34007297]

[38]   Mathlouthi N, Bouzaienne T, Oueslati I, *et al.* Use of rosemary, oregano, and a commercial blend of essential oils in broiler chickens: *In vitro* antimicrobial activities and effects on growth performance. J Anim Sci 2012; 90(3): 813-23.
[http://dx.doi.org/10.2527/jas.2010-3646] [PMID: 22064737]

[39]   Loetscher Y, Kreuzer M, Messikommer RE. Oxidative stability of the meat of broilers supplemented with rosemary leaves, rosehip fruits, chokeberry pomace, and entire nettle, and effects on performance and meat quality. Poult Sci 2013; 92(11): 2938-48.
[http://dx.doi.org/10.3382/ps.2013-03258] [PMID: 24135598]

[40]   Mahgoub SAM, El-Hack MEA, Saadeldin IM, Hussein MA, Swelum AA, Alagawany M. Impact of *Rosmarinus officinalis* cold-pressed oil on health, growth performance, intestinal bacterial populations, and immunocompetence of Japanese quail. Poult Sci 2019; 98(5): 2139-49.
[http://dx.doi.org/10.3382/ps/pey568] [PMID: 30590789]

[41]   Ghazalah AA, Ali AM. ali AM. Rosemary leaves as a dietary supplement for growth in broiler chickens. Int J Poult Sci 2008; 7(3): 234-9.
[http://dx.doi.org/10.3923/ijps.2008.234.239]

[42]   Petričević V, Lukić M, Škrbić Z, *et al.* The effect of using rosemary (*Rosmarinus officinalis*) in broiler nutrition on production parameters, slaughter characteristics, and gut microbiological population. Turk J Vet Anim Sci 2018; 42(6): 658-64.
[http://dx.doi.org/10.3906/vet-1803-53]

[43]   Hosseinzadeh S, Shariatmadari F, Karimi Torshizi MA, Ahmadi H, Scholey D. *Plectranthus amboinicus* and rosemary (*Rosmarinus officinalis* L.) essential oils effects on performance, antioxidant activity, intestinal health, immune response, and plasma biochemistry in broiler chickens. Food Sci Nutr 2023; 11(7): 3939-48.
[http://dx.doi.org/10.1002/fsn3.3380] [PMID: 37457190]

[44]   Nameghi AH, Edalatian O, Bakhshalinejad R. A blend of thyme and rosemary powders with poultry by-product meal can be used as a natural antioxidant in broilers. Acta Sci Anim Sci 2022; 45: e57126.
[http://dx.doi.org/10.4025/actascianimsci.v44i1.57126]

[45]   Yao Y, Liu Y, Li C, *et al.* Effects of rosemary extract supplementation in feed on growth performance, meat quality, serum biochemistry, antioxidant capacity, and immune function of meat ducks. Poult Sci 2023; 102(2): 102357.
[http://dx.doi.org/10.1016/j.psj.2022.102357] [PMID: 36502565]

[46]   Manafi M, Hedati M, Yari M. Effectiveness of rosemary (*Rosmarinus officinalis* l.) Essence on performance and immune parameters of broilers during aflatoxicosis. Adv Life Sci 2014; 4: 166-73.

[47]   Bölükbaşi Ş, Erhan M, Kaynar Ö. The effect of feeding thyme, sage and rosemary oil on laying hen performance, cholesterol and some proteins ratio of egg yolk and *Escherichia coli* count in feces. Arch. Geflügelk 2008; 72: 231-7.

[48]   Garcia ERDM, Chaves NRB, Oliveira CALD, Kiefer C, Melo EPD. Performance and egg quality of laying hens fed with mineral sources and rosemary oil. An Acad Bras Cienc 2019; 91(2): e20180516.
[http://dx.doi.org/10.1590/0001-3765201820180516] [PMID: 30758393]

[49]   Kedir S, Tamiru M, Tadese DA, *et al.* Effect of rosemary (*Rosmarinus officinalis*) leaf meal supplementation on production performance and egg quality of laying hens. Heliyon 2023; 9(8): e19124.
[http://dx.doi.org/10.1016/j.heliyon.2023.e19124]

[50]   Yang SS, Chen XY, Su AK. Effect of dietary supplementation with rosemary complex powder on the growth performance of native chickens. Braz. J. Poult. Sci 2023; 25: eRBCA-2022-1672.
[http://dx.doi.org/10.1590/1806-9061-2022-1672]

[51]   Elazab MA, Khalifah AM, Elokil AA, *et al.* Effect of dietary rosemary and ginger essential oils on the

growth performance, feed utilization, meat nutritive value, blood biochemicals, and redox status of growing NZW rabbits. Animals (Basel) 2022; 12(3): 375.
[http://dx.doi.org/10.3390/ani12030375] [PMID: 35158698]

[52] Attia YA, Hamed RS, Bovera F, *et al.* Milk thistle seeds and rosemary leaves as rabbit growth promoters. Anim. Sci 2019; Pap. Rep. 37:277-295.

[53] Mousa Z, Daghash H, Azoz AB, Mousa M, Farghaly M. Productive and physiological effects of chamomile and rosemary aqueous extract on New Zealand White growing rabbits. Egypt. J Rabbit Sci 2021; 31: 217-35.
[http://dx.doi.org/10.21608/ejrs.2021.211922]

[54] Naiel MAE, Ismael NEM, Shehata SA. Ameliorative effect of diets supplemented with rosemary (*Rosmarinus officinalis*) on aflatoxin B1 toxicity in terms of the performance, liver histopathology, immunity and antioxidant activity of Nile tilapia (*Oreochromis niloticus*). Aquaculture 2019; 511: 734264.
[http://dx.doi.org/10.1016/j.aquaculture.2019.734264]

[55] Naiel MAE, Ismael NEM, Negm SS, Ayyat MS, Al-Sagheer AA. Rosemary leaf powder–supplemented diet enhances performance, antioxidant properties, immune status, and resistance against bacterial diseases in Nile tilapia (*Oreochromis niloticus*). Aquaculture 2020; 526: 735370.
[http://dx.doi.org/10.1016/j.aquaculture.2020.735370]

[56] Kong F, Wang S, Dai D, *et al.* Preliminary investigation of the effects of rosemary extracts supplementation on milk production and rumen fermentation in high-producing dairy cows. Antioxidants 2022; 11(9): 1715.
[http://dx.doi.org/10.3390/antiox11091715] [PMID: 36139788]

[57] Souza KA, Monteschio JO, Mottin C, *et al.* Effects of diet supplementation with clove and rosemary essential oils and protected oils (eugenol, thymol and vanillin) on animal performance, carcass characteristics, digestibility, and ingestive behavior activities for Nellore heifers finished in feedlot. Livest Sci 2019; 220: 190-5.
[http://dx.doi.org/10.1016/j.livsci.2018.12.026]

[58] Yagoubi Y, Mekki I, Nasraoui M, Abdelmalek YB, Atti N. Effects of rosemary (*Rosmarinus officinalis*) by-products and linseed (*Linumusitatissimum*) intake on digestibility, body weight gain, and estimated tissular composition in cull fat-tailed ewes. Trop Anim Health Prod 2022; 54(2): 101.
[http://dx.doi.org/10.1007/s11250-022-03099-6] [PMID: 35146584]

[59] Farghaly M, Abdullah M. Effect of oregano, rosemary and peppermint as feed additives on nutrients digestibility, rumen fermentation and performance of fattening sheep. Egypt J Nutr Feeds 2021; 24(3): 365-76.
[http://dx.doi.org/10.21608/ejnf.2021.210838]

[60] Tawfeeq JA, Al-Omrani HA, Shaker RM, Hamza ZR, Abbas SF, Jabbar RH. Effect of peppermint and rosemary extractions on ruminant *in-vitro* digestibility. Adv Anim Vet Sci 2019; 7(10): 910-3.
[http://dx.doi.org/10.17582/journal.aavs/2019/7.10.910.913]

[61] Janz JAM, Morel PCH, Wilkinson BHP, Purchas RW. Preliminary investigation of the effects of low-level dietary inclusion of fragrant essential oils and oleoresins on pig performance and pork quality. Meat Sci 2007; 75(2): 350-5.
[http://dx.doi.org/10.1016/j.meatsci.2006.06.027] [PMID: 22063669]

[62] Di Turi L, Ragni M, Jambrenghi AC, *et al.* Effect of dietary rosemary oil on growth performance and flesh quality of farmed seabass (*Dicentrarchus labrax*). Ital J Anim Sci 2009; 8(sup2): 857-9.
[http://dx.doi.org/10.4081/ijas.2009.s2.857]

[63] Ayaz NO. Antidiabetic and renoprotective effects of water extract of *Rosmarinus officinalis* in streptozotocin-induced diabetic rat. Afr J Pharm Pharmacol 2012; 6: 2664-9.

[64] Afonso MS, de O Silva AM, Carvalho EBT, *et al.* Phenolic compounds from rosemary (*Rosmarinus*

officinalis L.) attenuate oxidative stress and reduce blood cholesterol concentrations in diet-induced hypercholesterolemic rats. Nutr Metab (Lond) 2013; 10(1): 19.
[http://dx.doi.org/10.1186/1743-7075-10-19] [PMID: 23374457]

[65] El-Gogary MR, El-Said EA, Mansour AM. Physiological and immunological effects of rosemary essential oil in growing rabbit diets. J Agric Sci 2018; 10: 485-91.
[http://dx.doi.org/10.5539/jas.v10n7p485]

[66] El-Gindy YM, Zahran SM, Ahmed MAR, Salem AZM, Misbah TR. Influence of dietary supplementation of clove and rosemary essential oils or their combination on growth performance, immunity status, and blood antioxidant of growing rabbits. Trop Anim Health Prod 2021; 53(5): 482.
[http://dx.doi.org/10.1007/s11250-021-02906-w] [PMID: 34562165]

[67] Shokrollahi B, Amini F, Fakour S, Andi MA. Effect of rosemary (*Rosmarinus Officinalis*) extract on weight, hematology and cell-mediated immune response of newborn goat kids. J Agric Rural Dev 2015; 116: 91-7.

[68] Alagawany M, Abd El-Hack M. The effect of rosemary herb as a dietary supplement on performance, egg quality, serum biochemical parameters, and oxidative status in laying hens. J Anim Feed Sci 2015; 24(4): 341-7.
[http://dx.doi.org/10.22358/jafs/65617/2015]

[69] Al-Shuwaili M. The effect of adding *Rosemarinus officinalis* and *Thymus vulgaris* to broilers diet on immune response and some physiological parameters of broilers. J Univ Kerbala 2014; 12: 92-7.

[70] Attia FM. Effects of dietary rosemary leaves and black seed on broiler performance. Egyptian Poultry Science Journal 2018; 38(2): 465-81.

[71] Lopez-Bote CJ, Gray JI, Gomaa EA, Flegal CJ. Effect of dietary administration of oil extracts from rosemary and sage on lipid oxidation in broiler meat. Br Poult Sci 1998; 39(2): 235-40.
[http://dx.doi.org/10.1080/00071669889187] [PMID: 9649877]

[72] Jridi M, Siala R, Fakhfakh N, *et al*. Effect of rosemary leaves and essential oil on turkey sausage quality. Acta Aliment 2015; 44(4): 534-41.
[http://dx.doi.org/10.1556/066.2015.44.0025]

[73] Ortuño J, Serrano R, Jordán MJ, Bañón S. Shelf life of meat from lambs given essential oil-free rosemary extract containing carnosic acid plus carnosol at 200 or 400mgkg−1. Meat Sci 2014; 96(4): 1452-9.
[http://dx.doi.org/10.1016/j.meatsci.2013.11.021] [PMID: 24412737]

[74] Perlo F, Fabre R, Bonato P, Jenko C, Tisocco O, Teira G. Refrigerated storage of pork meat sprayed with rosemary extract and ascorbic acid. Cienc Rural 2018; 48(4): e20170238.
[http://dx.doi.org/10.1590/0103-8478cr20170238]

[75] Liotta L, Chiofalo V, D'Alessandro E, Lo Presti V, Chiofalo B. Supplementation of rosemary extract in the diet of Nero Siciliano pigs: evaluation of the antioxidant properties on meat quality. Animal 2015; 9(6): 1065-72.
[http://dx.doi.org/10.1017/S1751731115000348] [PMID: 25997531]

[76] Tu Z, Moss-Pierce T, Ford P, Jiang TA. Rosemary (*Rosmarinus officinalis* L.) extract regulates glucose and lipid metabolism by activating AMPK and PPAR pathways in HepG2 cells. J Agric Food Chem 2013; 61(11): 2803-10.
[http://dx.doi.org/10.1021/jf400298c] [PMID: 23432097]

[77] Naimi M, Vlavcheski F, Shamshoum H, Tsiani E. Rosemary extract as a potential anti-hyperglycemic agent: current evidence and future perspectives. Nutrients 2017; 9(9): 968.
[http://dx.doi.org/10.3390/nu9090968] [PMID: 28862678]

[78] El-Huneidi W, Anjum S, Saleh MA, Bustanji Y, Abu-Gharbieh E, Taneera J. Carnosic acid protects INS-1 β-cells against streptozotocin-induced damage by inhibiting apoptosis and improving insulin secretion and glucose uptake. Molecules 2022; 27(7): 2102.

[http://dx.doi.org/10.3390/molecules27072102] [PMID: 35408495]

[79] Ibarra A, Cases J, Roller M, Chiralt-Boix A, Coussaert A, Ripoll C. Carnosic acid-rich rosemary (*Rosmarinus officinalis* L.) leaf extract limits weight gain and improves cholesterol levels and glycaemia in mice on a high-fat diet. Br J Nutr 2011; 106(8): 1182-9.
[http://dx.doi.org/10.1017/S0007114511001620] [PMID: 21676274]

[80] Bustanji Y, Issa A, Mohammad M, *et al.* Inhibition of hormone sensitive lipase and pancreatic lipase by *Rosmarinus officinalis* extract and selected phenolic constituents. J Med Plants Res 2010; 4: 2235-42.
[http://dx.doi.org/10.5897/JMPR10.399]

[81] Mann E, Sunni M, Bellin MD. Secretion of insulin in response to diet and hormones Pancreapdia. Exocrine Pancreas Knowledge Base 2020.
[http://dx.doi.org/10.3998/panc.2020.16]

[82] Al-Jamal AR, Alqadi T. Effects of rosemary (*Rosmarinus officinalis*) on lipid profile of diabetic rats. Jordan J Biol Sci 2011; 4: 199-204.

[83] Al Sheyab FM, Abuharfeil N, Salloum L, Bani Hani R, Awad DS. The effect of rosemary (*Rosmarinus officinalis* L.) plant extracts on the immune response and lipid profile in mice. J Biol Life Sci 2012; 3(1): 37-58.
[http://dx.doi.org/10.5296/jbls.v3i1.906]

[84] morsi RM, Mansour DS, Mousa AM. Ameliorative potential role of *Rosmarinus officinalis* extract on toxicity induced by etoposide in male albino rats. Braz J Biol 2024; 84: e258234.
[http://dx.doi.org/10.1590/1519-6984.258234] [PMID: 35830129]

[85] Abdel-Azeem AS, Hegazy AM, Zeidan HM, Ibrahim KS, El-Sayed EM. Potential renoprotective effects of rosemary and thyme against gentamicina toxicity in rats. J Diet Suppl 2017; 14(4): 380-94.
[http://dx.doi.org/10.1080/19390211.2016.1253632] [PMID: 27973970]

[86] El-Desouky MA, Mahmoud MH, Riad BY, Taha YM. Nephroprotective effect of green tea, rosmarinic acid and rosemary on N-diethylnitrosamine initiated and ferric nitrilotriacetate promoted acute renal toxicity in Wistar rats. Interdiscip Toxicol 2019; 12(2): 98-110.
[http://dx.doi.org/10.2478/intox-2019-0012] [PMID: 32206031]

[87] Wang H, Cheng J, Yang S, Cui SW, Wang M, Hao W. Rosemary extract reverses oxidative stress through activation of Nrf2 signaling pathway in hamsters fed on high fat diet and HepG2 cells. J Funct Foods 2020; 74: 104136.
[http://dx.doi.org/10.1016/j.jff.2020.104136]

[88] Rasoolijazi H, Mehdizadeh M, Soleimani M, Nikbakhte F, Eslami Farsani M, Ababzadeh S. The effect of rosemary extract on spatial memory, learning and antioxidant enzymes activities in the hippocampus of middle-aged rats. Med J Islam Repub Iran 2015; 29: 187.
[PMID: 26034740]

[89] Liu Y, Li C, Huang X, *et al.* Dietary rosemary extract modulated gut microbiota and influenced the growth, meat quality, serum biochemistry, antioxidant, and immune capacities of broilers. Front Microbiol 2022; 13: 1024682.
[http://dx.doi.org/10.3389/fmicb.2022.1024682] [PMID: 36338103]

[90] Rostami H, Seidavi A, Dadashbeiki M, Asadpour Y, Simões J. Effects of different dietary *Rosmarinus officinalis* powder and vitamin E levels on the performance and gut gross morphometry of broiler chickens. Rev Bras Cienc Avic 2015; 17(spe): 23-30.
[http://dx.doi.org/10.1590/1516-635XSPECIALISSUENutrition-PoultryFeedingAdditives023-030]

[91] Mahfuz S, Shang Q, Piao X. Phenolic compounds as natural feed additives in poultry and swine diets: a review. J Anim Sci Biotechnol 2021; 12(1): 48.
[http://dx.doi.org/10.1186/s40104-021-00565-3] [PMID: 33823919]

[92] Kumar D, Arya V, Kaur R, Bhat ZA, Gupta VK, Kumar V. A review of immunomodulators in the

Indian traditional health care system. J. Microbiol 2011; Immunol. Infect. 45: 165-184.
[http://dx.doi.org/10.1016/j.jmii.2011.09.030]

[93] Hashemipour H, Kermanshahi H, Golian A, Veldkamp T. Effect of thymol and carvacrol feed supplementation on performance, antioxidant enzyme activities, fatty acid composition, digestive enzyme activities, and immune response in broiler chickens. Poult Sci 2013; 92(8): 2059-69.
[http://dx.doi.org/10.3382/ps.2012-02685] [PMID: 23873553]

[94] Soltani M, Tabeidian SA, Ghalamkari G, Adeljoo AH, Mohammadrezaei M, Fosoul SSAS. Effect of dietary extract and dried areal parts of *Rosmarinus officinalis* on performance, immune responses and total serum antioxidant activity in broiler chicks. Asian Pac J Trop Dis 2016; 6(3): 218-22.
[http://dx.doi.org/10.1016/S2222-1808(15)61017-9]

[95] Labban L, Mustafa UE-S, Ibrahim YM. The effects of rosemary (*Rosmarinus officinalis*) leaves powder on glucose level, lipid profile and lipid peroxidation. Int J Clin Med 2014; 5(6): 297-304.
[http://dx.doi.org/10.4236/ijcm.2014.56044]

[96] Aghamiri SM, Eslami Farsani M, Seyedebrahimi R, Sarikhani MJ, Ababzadeh S. Synergic effects of rosemary extract and aerobic exercise on sperm parameters and testicular tissue in an aged rat model. Gene Cell Tissue 2023; 10(3): e130832.
[http://dx.doi.org/10.5812/gct-130832]

[97] Nusier MK, Bataineh HN, Daradkah HM. Adverse effects of rosemary (*Rosmarinus officinalis* L.) on reproductive function in adult male rats. Exp Biol Med (Maywood) 2007; 232(6): 809-13.
[http://dx.doi.org/10.3181/00379727-232-2320809] [PMID: 17526773]

[98] Hamza FZ, Al-Sharafi NM, Kasim SF. Effect of aqueous rosemary extract on some sexual hormones in male rats with high thyroxine level. Iraqi J Vet Sci 2021; 35(2): 369-73.
[http://dx.doi.org/10.33899/ijvs.2020.126872.1404]

[99] Hasan AS, Al-Rikaby AA. Evaluating the influence of rosemary leaves extract on hormonal and histopathological alterations in male rabbits exposed to cypermethrin. Arch Razi Inst 2023; 78(3): 823-31.
[PMID: 38028834]

[100] Heidari-Vala H, Ebrahimi Hariry R, Sadeghi MR, Akhondi MM, Ghaffari Novin M, Heidari M. Evaluation of an aqueous-ethanolic extract from *Rosmarinus officinalis* (rosemary) for its activity on the hormonal and cellular function of testis in adult male rat. Iran J Pharm Res 2013; 12(2): 445-51.
[PMID: 24250620]

[101] Chiofalo V, Liotta L, Fiumanò R, Riolo EB, Chiofalo B. Influence of dietary supplementation of *Rosmarinus officinalis* L. on performances of dairy ewes organically managed. Small Rumin Res 2012; 104(1-3): 122-8.
[http://dx.doi.org/10.1016/j.smallrumres.2011.09.051]

[102] Smeti S, Hajji H, Bouzid K, *et al.* Effects of *Rosmarinus officinalis* L. as essential oils or in form of leaves supplementation on goat's production and metabolic statute. Trop Anim Health Prod 2015; 47(2): 451-7.
[http://dx.doi.org/10.1007/s11250-014-0721-3] [PMID: 25425356]

[103] Kholif AE, Matloup OH, Morsy TA, *et al.* Rosemary and lemongrass herbs as phytogenic feed additives to improve efficient feed utilization, manipulate rumen fermentation and elevate milk production of Damascus goats. Livest Sci 2017; 204: 39-46.
[http://dx.doi.org/10.1016/j.livsci.2017.08.001]

Silybum marianum (Milk Thistle)

Youssef A. Attia[1,2,*], **Mohamed E. Abd El-Hack**[3], **Mahmoud M. Alagawany**[3], **Rashed A. Alhotan**[4], **Salem R. Alyileili**[5], **Hossam A. Shahba**[6], **Asmaa F. Khafaga**[7] and **Maria Cristina de Oliveira**[8]

[1] *Sustainable Agriculture Production Research Group, Agriculture Department, Faculty of Environmental Sciences, King Abdulaziz University, Jeddah-21589, Saudi Arabia*

[2] *Animal and Poultry Production Department, Faculty of Agriculture, Damanhour University, Damanhour-22713, Egypt*

[3] *Poultry Department, Faculty of Agriculture, Zagazig University, Zagazig-44519, Egypt*

[4] *Department of Animal Production, College of Food and Agricultural Sciences, King Saud University, Riyadh, Saudi Arabia*

[5] *Department of Laboratory Analyses, College of Food and Agriculture Sciences, United Arab Emirates University, AlAin United Arab Emirates*

[6] *Rabbit, Turkey and Water Fowl Research Department, Animal Production Research Institute, Agriculture Research Center, Dokki, Giza-3751310, Egypt*

[7] *Department of Pathology, Faculty of Veterinary Medicine, Alexandria University, Apis, Alexandria, 21944, Egypt*

[8] *University of Rio Verde, Faculty of Veterinary Medicine, Rio Verde, GO, Brazil*

Abstract: Silymarin, a polyphenolic flavonoid complex extracted from milk thistle seeds (*Silybum marianum*), has a wide range of therapeutic properties, including anti-inflammatory, immunomodulatory, and antioxidant effects. This review explores the applications of milk thistle and silymarin in animal nutrition, focusing on their effects on productive performance, animal health, metabolic profiles, and detoxification processes. Milk thistle seeds contain various nutritional components that have been shown to improve nutrient utilization, stimulate appetite, and enhance the intestinal environment. Studies have reported that milk thistle supplementation significantly improves productive performance, carcass yield, and digestibility in growing rabbits, broilers, and quail. The hepatoprotective effects of milk thistle are attributed to its inhibition of lipid peroxidation, stabilization of membrane permeability, reduction of apoptosis in hepatocytes, and limited leakage of hepatic enzymes. The potent antioxidant properties of silymarin protect cells from oxidative stress by scavenging reactive oxygen species and inhibiting lipid peroxidation. Milk thistle extract also acts

* **Corresponding author Youssef A. Attia:** Sustainable Agriculture Production Research Group, Agriculture Department, Faculty of Environmental Sciences, King Abdulaziz University, Jeddah-21589, Saudi Arabia & Animal and Poultry Production Department, Faculty of Agriculture, Damanhour University, Damanhour-22516, Egypt; E-mail: yaattia@kau.edu.sa

as a free radical scavenger, protecting against glutathione depletion, and enhancing glutathione peroxidase activity in the brain and kidneys. Furthermore, milk thistle supplementation has been shown to improve hematological parameters, such as leukocyte count, hemoglobin levels, and packed cell volume, in birds exposed to ochratoxin A. The beneficial effects of milk thistle on animal immunity, oxidative stress, performance, and reproduction make it a valuable candidate for use as a feed additive in animal nutrition.

Keywords: Antioxidants, Livestock, Liver function, Milk thistle, Productive performance, Silymarin.

INTRODUCTION

Silybum marianum, commonly known as milk thistle, is a spiny herb belonging to the *Asteraceae* family and has a long history of medicinal use. Theophrastus, in the 4th century BC, was likely the first to describe it under the name "Pternix." Later, it was referenced by Dioskurides in Materia Medica and by Pliny the Elder in the 1st century AD [1]. Milk thistle seeds (MTS) contain various nutritional components, with moisture levels ranging from 5.01% to 6.27%, ash from 1.25% to 2.37%, fat from 19.74% to 23.19%, fiber from 4.39% to 7.40%, protein from 20.74% to 30.09%, and nitrogen-free extract from 34.13% to 45.42% [2].

Silymarin, a polyphenolic flavonoid complex extracted from MTS [3], comprises silybin A and B, isosilybin A and B, silydianin, silychristin, and isosilychristin [4 - 7]. These compounds have demonstrated estrogenic effects in ovariectomized rats [8], with silibinin binding to cytosolic estrogen receptors [9, 10]. The antioxidant properties of milk thistle have been extensively studied and confirmed by Camini and Costa [5], Aghemo *et al.* [11], and Yardimci *et al.* [12], with silybin being the primary bioactive component, constituting about 4-6% of the milk thistle seed extract [13].

In addition to silybin, MTS contains other health-promoting components, including lipids such as triglycerides, proteins, sugars, tocopherol, sterols [14], silybonol, apigenin, betaine, fixed oils, and free fatty acids [15]. Milk thistle has been used for centuries in Europe to treat various dysfunctions, particularly in the liver, gall bladder, heart, and kidneys [16 - 20]. The health effects of silymarin are attributed to its numerous beneficial properties, including anti-inflammatory [21, 22], immunomodulatory [23, 24], and antioxidant activities [5, 11]. Additionally, silymarin enhances protein synthesis [6, 25] and mitigates the effects of toxins [26, 27]. The active ingredients present in MTS and their beneficial properties are shown in Fig. (**1**).

Fig. (1). Active ingredients present in MTS.

Effect of Milk Thistle on Productive Performance

Milk thistle seeds (MTS) have demonstrated the potential to improve nutrient utilization [28, 29], stimulate appetite [30, 31], and enhance the intestinal environment by increasing lactic acid bacteria concentrations and reducing harmful bacteria [32, 33]. These effects positively influence the productive performance of the animals.

Several studies have reported the effect of milk thistle on the productive performance of poultry and other animals [10, 25, 34]. For instance, Attia *et al.* [33] observed that MTS supplementation at 10 g/kg significantly improved productive performance, carcass yield, and digestibility of crude protein, organic matter, and dry matter in growing rabbits compared with the control group, with the best results at 10 g/kg inclusion. Similarly, Bagno *et al.* [35] demonstrated that milk thistle extracts enhanced anabolic processes and increased the protein utilization of the albumin fraction for organogenesis in broilers. In quails, Youssef *et al.* [26] found that dietary supplementation with curcumin and silymarin at 250 mg/kg significantly improved the growth rate compared to an unsupplemented diet. The effect was further enhanced by 500 mg/kg of curcumin and silymarin, although no significant changes were observed in feed intake, feed conversion, carcass traits, or blood protein components.

Recent research by Elnesr *et al.* [10] highlighted that milk thistle in poultry nutrition enhances productivity and growth performance. The hepatoprotective effects of MTS are linked to its inhibition of lipid peroxidation and stabilization of membrane permeability [19]. MTS may also reduce apoptosis in hepatocytes and limit the leakage of hepatic enzymes such as ALT and AST [36]. Ilyas *et al.* [37] reported similar findings, noting that MTS supplementation decreased ALT, AST, alkaline phosphatase (ALP), and total bilirubin levels, keeping them within normal ranges compared to the hepatotoxic group. Additionally, Ahmad *et al.* [38] found that birds fed silymarin, vitamin E, or their combination exhibited similar levels of total protein, albumin, ALT, creatinine, and urea as the control group, while the combination of these additives improved physiological responses in birds exposed to 1000 µg of ochratoxin A.

Silymarin is recognized for its potent antioxidant properties, scavenging reactive oxygen species (ROS), and inhibiting lipid peroxidation, which protects cells from oxidative stress [7, 39, 40]. Ramadan *et al.* [41] revealed that repeated oral doses of milk thistle extract significantly decreased liver enzyme activity while increasing antioxidant enzyme levels in rat liver homogenate, highlighting its role as a powerful free radical scavenger. The effects of MTS on animal immunity, oxidative stress, performance, and reproduction are summarized in Fig. (**2**).

Effect of Milk Thistle on Blood Profile

Milk thistle extract (MTE) acts as a potent free radical scavenger, protecting against glutathione (GSH) depletion and oxidative stress [42] and enhancing glutathione peroxidase activity in both the brain and kidneys [43]. Oxidative stress is known to reduce the blood cell count and hemoglobin (Hb) levels. Ahmad *et al.* [38] demonstrated that birds exposed to 1000 µg of ochratoxin A, when supplemented with silymarin and vitamin E, exhibited hematobiochemical profiles similar to those of the control group. These birds showed improved hematological parameters, such as leukocyte count, hemoglobin levels, and packed cell volume. Additionally, Attia *et al.* [34] reported that supplementing diets with 10 g/kg MTS significantly enhanced red blood cell and lymphocyte counts and improved liver function markers and total antioxidant capacity. MTS also induced moderate activation of the lymphoid follicles in the spleen at this concentration.

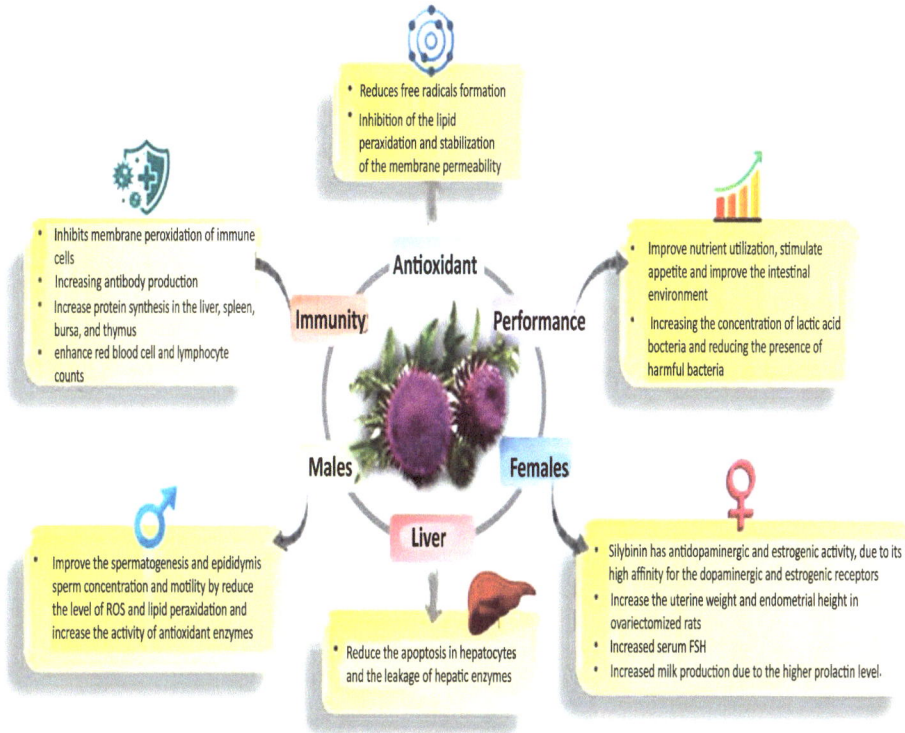

Fig. (2). Effects of MTS on animal immunity, oxidation, performance, and reproduction.

El-Adawi *et al.* [44] found that supplementation with MTS, grape seed extract, or a combination of both, increased glutathione peroxidase levels by 76%, 76%, and 35%, respectively, while reducing malondialdehyde (MDA) levels by 47%, 42%, and 29%. These treatments also decreased creatinine concentrations by 16%, 15%, and 2%, respectively. Lower levels of free radicals were associated with decreased MDA and increased GSH levels, suggesting that antioxidant enzymes are less extensively used to combat oxidative stress. These results align with those of Abdalla *et al.* [45] and Khazaei *et al.* [46], who studied broilers and laboratory animals, respectively.

Sabiu *et al.* [47] demonstrated that oral administration of 400 mg/kg body weight of acetaminophen, a known hepatotoxic agent, over seven days resulted in a significant elevation of serum alkaline phosphatase (ALP), alanine transaminase (ALT), aspartate transaminase (AST), bilirubin, and total protein in rats. However, supplementation with silymarin, vitamin C, or a combination of both normalized these levels. Additionally, silymarin and vitamin C reduced thiobarbituric acid-reactive substances, indicating reduced lipid peroxidation, and restored the activity of antioxidant enzymes, such as superoxide dismutase, catalase, glutathione peroxidase, and glutathione-s-transferase in hepatotoxic rats, further confirming the antioxidant potential of MTS.

Furthermore, Gobalakrishnan *et al.* [48] observed that silymarin supplementation increased bile acid excretion in rats, leading to depletion of the bile acid pool and increased bile acid synthesis. Silymarin also improved the binding of low-density lipoprotein (LDL) to hepatocytes, thereby reducing plasma LDL levels [49].

Effect of Milk Thistle on Immunity

Phytogenic substances such as milk thistle have been widely recognized for their immunomodulatory effects on poultry, enhancing overall health and performance without adverse effects [50, 51]. Silymarin, the active compound in milk thistle, plays a crucial role in reducing the formation of free radicals and inhibiting the peroxidation of immune cell membranes [52]. This antioxidant action is essential for improving immune responses, particularly by increasing antibody production. Silymarin also supports protein synthesis in key immune organs such as the liver, spleen, bursa, and thymus, leading to stronger immune responses. Chand *et al.* [53] and Makki *et al.* [54] demonstrated that silymarin supplementation resulted in higher antibody titers against important poultry diseases such as Newcastle Disease Virus (NDV) and Infectious Bursal Disease (IBD). This improvement in immunity was linked to the potent antioxidant effects of silymarin, which reduces oxidative stress in immune cells. Additionally, milk thistle supplementation has been shown to increase the relative weight and improve the morphology of the bursa, a key immune organ in birds, especially in broilers fed aflatoxin-contaminated diets [55, 56]. The positive effect of milk thistle on animal immunity is largely due to its ability to scavenge free radicals and preserve the body's glutathione supply, which protects immune cells from damage [5, 11]. Silymarin ensures better immune competence by safeguarding immune system resources, and the importance of the bursa and thymus in immunological competence has been well established. Surgical or hormonal bursectomy in neonatal or embryonic poultry has been shown to drastically reduce antibody production, highlighting the critical role of these organs in maintaining a healthy immune system [58]. The protective and supportive role of milk thistle in these organs further emphasizes its value in poultry nutrition.

Effect of Milk Thistle on Semen Quality and Fertility

Silibinin, a major component of silymarin, has shown potent antioxidant properties primarily through its ability to control free radical formation and inhibit lipid peroxidation in cell membranes [19]. This antioxidant action suggests that silibinin may significantly improve sperm quality and promote spermatogenesis in males [59]. Spermatozoa membranes, rich in polyunsaturated fatty acids, are particularly vulnerable to oxidative stress, which leads to the production of harmful compounds such as acrolein, malondialdehyde (MDA), and 4-hydroxynonenal [60, 61]. These compounds impair sperm viability and motility, but silymarin helps reduce reactive oxygen species (ROS) levels and lipid peroxidation, while enhancing antioxidant enzyme activity [62], protecting gonadal cells, and improving sperm quality.

For instance, Oufi *et al.* [63] observed that silibinin supplementation in mice led to a non-significant improvement in sperm motility and a reduction in dead sperm cells, although the results were not significant. In a separate study, rats receiving 151.2 mg/kg of silymarin for 30 days exhibited increased levels of testosterone and luteinizing hormone (LH) compared with the control group, with complete spermatogenesis observed in the seminiferous tubules [64]. These findings suggest that silymarin enhances reproductive health by supporting testosterone production and sperm development.

Other studies have also confirmed the positive effects of silibinin on spermatogenesis. Muthu and Prabu [65] reported improvements in sperm viability, concentration, and motility in rats, while Hamed *et al.* [66] noted better sperm quality, higher serum testosterone levels, and increased fertility rates in bucks fed diets containing 10 g/kg of milk thistle seeds (MTS). Similar beneficial effects were demonstrated by Kranti *et al.* [40] in rats and Abid Ali *et al.* [67] in rabbits, further supporting silymarin's role in promoting male reproductive health.

The benefits of silymarin extend to female reproductive health. Studies have shown that silibinin can bind to estrogen receptors [9] and increase uterine weight and endometrial height in ovariectomized rats after 30 days of supplementation [8]. Farmer *et al.* [68] found that gilts receiving 8 g/day of silymarin had higher prolactin levels at 94 days of gestation, although progesterone and estradiol levels remained unchanged. In female rats, silymarin supplementation for 30 to 60 days resulted in an increase in serum follicle-stimulating hormone (FSH) levels, with no significant changes in estradiol or LH levels. The ovaries of these rats also displayed many growing follicles, and hypertrophy of the uterine endometrial epithelium was observed [64]. Moreover, Refaie *et al.* [69] found that supplementing rabbit diets with ethanolic milk thistle extract improved litter weight at birth and weaning, which was attributed to the higher milk production driven by elevated prolactin levels. These findings underline the beneficial effects of silymarin on reproductive health, both in terms of fertility and overall reproductive organ function, in males and females.

Effect of Milk Thistle on Milk Yield

Silymarin has shown promising effects on lactation, primarily through its ability to increase prolactin, a hormone essential for milk production. Studies on rats and other animals have demonstrated this lactogenic effect. For example, the increased prolactin concentration in rats treated with silymarin is thought to result from its estrogenic and antidopaminergic properties [70 - 72]. As dopamine inhibits prolactin secretion, the action of silibinin on dopaminergic and estrogenic receptors plays a crucial role in elevating prolactin levels [73, 74]. The high affinity of silibinin for these receptors explains its dual impact of blocking dopamine and enhancing estrogenic effects [9, 75, 76].

In human studies, silymarin demonstrated a significant increase in milk production. Women who took 420 mg of silymarin daily for 63 days exhibited an 85.94% increase in daily milk production compared to only 32.09% in the placebo group, indicating its potent galactagogue (milk-promoting) effects [78]. These results align with the estrogenic activity of silymarin, which can positively influence lactation [77].

The effects of silymarin on lactation have also been observed in dairy animals. Tedesco *et al.* [79, 81] found that cows receiving milk thistle seed (MTS) supplementation at 10 g/day during early lactation experienced an increase in milk production. These cows reached their peak milk production a week earlier than the control group, and daily milk output increased by 5-6 liters per cow. This suggests a clear lactogenic effect of silymarin on cows, particularly during the early lactation period. Additionally, silymarin supplementation results in decreased beta-hydroxybutyric acid levels and increased ketonuria, which are indicators of enhanced metabolic function in lactating cows [50, 77].

However, while milk production increased, no significant changes were observed in milk composition or liver histology and biochemistry in cows receiving silymarin, indicating that its effects may be more focused on quantity rather than quality or health metrics [82, 79, 83].

In animal models such as female rats, silymarin supplementation induced ductal and lobuloalveolar hyperplasia, as evidenced by an increase in the number and size of alveoli and lobules, which are key structures involved in milk secretion

[71]. This morphological change in mammary tissues further supports the role of silymarin in promoting milk production and lactation.

CONCLUSION AND REMARKS

Milk thistle (MTS) has numerous beneficial properties that make it valuable for animal nutrition. Its anti-inflammatory and antioxidant effects help protect cells from oxidative stress and damage, whereas its appetite-stimulating capacity and ability to improve the intestinal microenvironment support better nutrient absorption and overall health. MTS also functions as an immunomodulator, enhancing the immune response by increasing antibody production and protecting the immune cells from oxidative stress. Furthermore, it increases prolactin secretion, which boosts milk production in lactating animals. Owing to these diverse properties, MTS can be incorporated into animal diets to improve productive performance, enhance the blood biochemical profile, boost immunity, and promote better semen quality and fertility. MTS can stimulate increased milk yield in lactating animals, offering significant benefits for dairy production. Thus, the inclusion of MTS in animal nutrition has the potential to optimize growth, reproduction, and overall health across various species.

IMPLICATIONS

Extensive research on milk thistle (*Silybum marianum*) and its active component, silymarin, suggests a significant potential for its application in animal nutrition. The diverse beneficial properties of milk thistle, including its antioxidant, anti-inflammatory, immunomodulatory, and hepatoprotective effects, imply that its incorporation into animal feed could lead to improved overall health, enhanced productive performance, and better reproductive outcomes in various livestock species. Furthermore, the demonstrated ability of silymarin to increase prolactin levels and stimulate milk production suggests that it could be particularly valuable for dairy animal nutrition, potentially leading to increased milk yields and earlier peak production in lactating animals. Collectively, these findings imply that milk thistle supplementation could be a promising strategy for optimizing animal health, productivity, and profitability in livestock farming, warranting further research and consideration for widespread implementation in animal nutrition programs.

REFERENCES

[1]   Křen V, Walterová D. Silybin and silymarin - new effects and applications. Biomed Pap Med Fac Univ Palacky Olomouc Czech Repub 2005; 149(1): 29-41.
[http://dx.doi.org/10.5507/bp.2005.002] [PMID: 16170386]

[2]   Aziz M, Saeed F, Ahmad N, *et al.* RETRACTED: Biochemical profile of milk thistle (*Silybum Marianum* L.) with special reference to silymarin content. Food Sci Nutr 2021; 9(1): 244-50.

[http://dx.doi.org/10.1002/fsn3.1990] [PMID: 33473288]

[3]   Zarenezhad E, Abdulabbas HT, Kareem AS, *et al.* Protective role of flavonoids quercetin and silymarin in the viral-associated inflammatory bowel disease: an updated review. Arch Microbiol 2023; 205(6): 252.
[http://dx.doi.org/10.1007/s00203-023-03590-0] [PMID: 37249707]

[4]   Šuk J, Jašprová J, Biedermann D, *et al.* Isolated silymarin flavonoids increase systemic and hepatic bilirubin concentrations and lower lipoperoxidation in mice. Oxid Med Cell Longev 2019; 2019: 1-12.
[http://dx.doi.org/10.1155/2019/6026902] [PMID: 30891115]

[5]   Camini FC, Costa DC. Silymarin: not just another antioxidant. J Basic Clin Physiol Pharmacol 2020; 31(4): 20190206.
[http://dx.doi.org/10.1515/jbcpp-2019-0206] [PMID: 32134732]

[6]   Vargas-Mendoza N, Morales-González Á, Morales-Martínez M, *et al.* Flavolignans from silymarin as Nrf2 bioactivators and their therapeutic applications. Biomedicines 2020; 8(5): 122.
[http://dx.doi.org/10.3390/biomedicines8050122] [PMID: 32423098]

[7]   Akhtar MN, Saeed R, Saeed F, *et al.* Silymarin: a review on paving the way towards promising pharmacological agent. Int J Food Prop 2023; 26(1): 2256-72.
[http://dx.doi.org/10.1080/10942912.2023.2244685]

[8]   Kummer V, Mašková J, Čanderle J, Zralý Z, Neča J, Machala M. Estrogenic effects of silymarin in ovariectomized rats. Vet Med (Praha) 2001; 46(1): 17-23.
[http://dx.doi.org/10.17221/7846-VETMED]

[9]   Dupuis ML, Conti F, Maselli A, *et al.* The natural agonist of estrogen receptor β, silibinin, plays an immunosuppressive role as a potential therapeutic tool in rheumatoid arthritis. Front Immunol 2018; 9: 1903.
[http://dx.doi.org/10.3389/fimmu.2018.01903] [PMID: 30174672]

[10]   Elnesr Shaaban S, Elwan Hamada AM. El Sabry, Mohamed I, Shehata, Abdelrazeq M. The nutritional importance of milk thistle (*Silybum* marianum) and its beneficial influence on poultry. Worlds Poult Sci J 2023; 79.
[http://dx.doi.org/10.1080/00439339.2023.2234339]

[11]   Aghemo A, Alekseeva OP, Angelico F, *et al.* Role of silymarin as antioxidant in clinical management of chronic liver diseases: a narrative review. Ann Med 2022; 54(1): 1548-60.
[http://dx.doi.org/10.1080/07853890.2022.2069854] [PMID: 35635048]

[12]   Yardımcı M, Göz M, Aydın MS, Kankılıç N, Temiz E. Antioxidant actions of thymoquinone, silymarin, and curcumin on experimental aortic ischemia-reperfusion model in Wistar albino rats. Rev Bras Cir Cardiovasc 2022; 37(6): 807-13.
[http://dx.doi.org/10.21470/1678-9741-2021-0462] [PMID: 35657313]

[13]   Greenlee H, Abascal K, Yarnell E, Ladas E. Clinical applications of *Silybum marianum* in oncology. Integr Cancer Ther 2007; 6(2): 158-65.
[http://dx.doi.org/10.1177/1534735407301727] [PMID: 17548794]

[14]   Abenavoli L, Capasso R, Milic N, Capasso F. Milk thistle in liver diseases: past, present, future. Phytother Res 2010; 24(10): 1423-32.
[http://dx.doi.org/10.1002/ptr.3207] [PMID: 20564545]

[15]   Marderosian AD, Beutler JA. The review of natural products: the most complete source of natural information 3. Alphen aan den Rijn 2002.

[16]   Rafieian-Kopaie M, Nasri H. Silymarin and diabetic nephropathy. J Renal Inj Prev 2012; 1(1): 3-5.
[http://dx.doi.org/10.12861%2Fjrip.2012.02] [PMID: 25340091]

[17]   Amiri M, Motamedi P, Vakili L, *et al.* Beyond the liver protective efficacy of silymarin; bright renoprotective effect on diabetic kidney disease. J Nephropharmacol 2014; 3(2): 25-6. Available from: [www.ncbi.nlm.nih.gov/pmc/articles/PMC5297522/pdf/npj-3-25.pdf].

[PMID: 28197456]

[18]   Abenavoli L, Izzo AA, Milić N, Cicala C, Santini A, Capasso R. Milk thistle (*Silybum marianum*): A concise overview on its chemistry, pharmacological, and nutraceutical uses in liver diseases. Phytother Res 2018; 32(11): 2202-13.
[http://dx.doi.org/10.1002/ptr.6171] [PMID: 30080294]

[19]   Gillessen A, Schmidt HHJ. Silymarin as supportive treatment in liver diseases: a narrative review. Adv Ther 2020; 37(4): 1279-301.
[http://dx.doi.org/10.1007/s12325-020-01251-y] [PMID: 32065376]

[20]   Zalat Z, Kohaf N, Alm El-Din M, Elewa H, Abdel-Latif M. Silymarin: A promising cardioprotective agent. Azhar International Journal of Pharmaceutical and Medical Sciences 2021; 1(1): 15-23.
[http://dx.doi.org/10.21608/aijpms.2021.52962.1014]

[21]   Ferraz AC, Almeida LT, da Silva Caetano CC, *et al.* Hepatoprotective, antioxidant, anti-inflammatory, and antiviral activities of silymarin against mayaro virus infection. Antiviral Res 2021; 194: 105168.
[http://dx.doi.org/10.1016/j.antiviral.2021.105168] [PMID: 34437912]

[22]   Zhao X, Wang H, Yang Y, *et al.* Protective effects of silymarin against D-Gal/LPS-induced organ damage and inflammation in mice. Drug Des Devel Ther 2021; 15: 1903-14.
[http://dx.doi.org/10.2147/DDDT.S305033] [PMID: 33976540]

[23]   Gharagozloo M, Jafari S, Esmaeil N, Javid EN, Bagherpour B, Rezaei A. Immunosuppressive effect of silymarin on mitogen-activated protein kinase signalling pathway: the impact on T cell proliferation and cytokine production. Basic Clin Pharmacol Toxicol 2013; 113(3): 209-14.
[http://dx.doi.org/10.1111/bcpt.12088] [PMID: 23701595]

[24]   Esmaeil N, Anaraki SB, Gharagozloo M, Moayedi B. Silymarin impacts on immune system as an immunomodulator: One key for many locks. Int Immunopharmacol 2017; 50: 194-201.
[http://dx.doi.org/10.1016/j.intimp.2017.06.030] [PMID: 28672215]

[25]   Achufusi TGO, Patel RK. Milk thistle. Treasure Island, FL: StatPearls Publishing 2022.

[26]   Youssef SF, Sayed-ElAhl RMH, Mohamed MHA, El-Gabry HE, Abd El-Halim HAH, Eshera AA. Supplementing growing quail diets with silymarin and curcumin to improve productive performance and antioxidant status and alleviate aflatoxin b1 adverse effects during the summer season. Egyptian Journal of Agricultural Research 2022; 100(4): 529-39.
[http://dx.doi.org/10.21608/ejar.2022.155457.1263]

[27]   Bencze-Nagy J, Strifler P, Horváth B, *et al.* Effects of dietary milk thistle (*Silybum marianum*) supplementation in ducks fed mycotoxin-contaminated diets. Vet Sci 2023; 10(2): 100.
[http://dx.doi.org/10.3390/vetsci10020100] [PMID: 36851404]

[28]   Faryadi S, Sheikhahmadi A, Farhadi A, Nourbakhsh H. Effects of silymarin and nano-silymarin on performance, egg quality, nutrient digestibility, and intestinal morphology of laying hens during storage. Ital J Anim Sci 2021; 20(1): 1633-44.
[http://dx.doi.org/10.1080/1828051X.2021.1975503]

[29]   Shanmugam S, Park JH, Cho S, Kim IH. Silymarin seed extract supplementation enhances the growth performance, meat quality, and nutrients digestibility, and reduces gas emission in broilers. Anim Biosci 2022; 35(8): 1215-22.
[http://dx.doi.org/10.5713/ab.21.0539] [PMID: 35240024]

[30]   Bendowski W, Michalczuk M, Jóźwik A, *et al.* Using milk thistle (*Silybum marianum*) extract to improve the welfare, growth performance and meat quality of broiler chicken. Animals (Basel) 2022; 12(9): 1085.
[http://dx.doi.org/10.3390/ani12091085] [PMID: 35565511]

[31]   Reisi N, Esmaeil N, Gharagozloo M, Moayedi B. Therapeutic potential of silymarin as a natural iron-chelating agent in β-thalassemia intermedia. Clin Case Rep 2022; 10(1): e05293.
[http://dx.doi.org/10.1002/ccr3.5293] [PMID: 35106163]

[32]   Bhowmik D, Dubey J, Mehra S. Probiotic efficiency of *Spirulina platensis*-stimulating growth of lactic acid bacteria. Am-Eurasian J Agric Environ Sci 2009; 6: 546-9. Available from: [www.idosi.org/aejaes/jaes6(5)/9.pdf].

[33]   Bessam FH, Mehdadi Z. Evaluation of the antibacterial and antifongigal activity of different extract of flavonoiques *Silybum marianum* L. Adv Environ Biol 2014; 8: 17. [http://www.aensiweb.net/AENSIWEB/aeb/aeb/September%202014/1-9.pdf].

[34]   Attia YA, Hamed RS, Bovera F, Al-Harthi MA. Abd El-Hamid AElH, Esposito L, Shahba HA. Milk thistle seeds and rosemary leaves as rabbit growth promoters. Anim Sci Pap Rep 2019; 37: 277-95. [http://www.igbzpan.pl/uploaded/FSiBundleContentBlockBundleEntityTranslatableBlockTranslatable FilesElement/filePath/1459/277-296.pdf].

[35]   Bagno O, Shevchenko S, Shevchenko A, *et al.* Physiological status of broiler chickens with diets supplemented with milk thistle extract. Vet World 2021; 14(5): 1319-23. [http://dx.doi.org/10.14202/vetworld.2021.1319-1323] [PMID: 34220137]

[36]   Karimi G, Vahabzadeh M, Lari P, Rashedinia M, Moshiri. M. Iran J Med Sci 2011; 14: 308-17. Available from: [www.ncbi.nlm.nih.gov/pmc/articles/PMC3586829/pdf/IJBMS-14-308.pdf].

[37]   Ilyas N, Sadiq M Jehangir A. Hepatoprotective effect of garlic (*Allium Sativum*) and milk thistle (silymarin) in isoniazid induced hepatotoxicity in rats biomedical 2011; 27: 166-170. Available from: https://citeseerx.ist.psu.edu/document?repid=rep1&type=pdf&doi=4ff7c8bc06f9b91f4d8a3b89a90e1e 432f7d22bf

[38]   Ahmad D, Saleemi K, Khan Z, *et al.* Effects of ochratoxin A feeding in white leghorn cockerels on hematological and serum biochemical parameters and its amelioration with silymarin and vitamin E. Pak Vet J 2012; 32: 520-4. [http://www.pvj.com.pk/pdf-files/32_4/520-524.pdf].

[39]   Suksomboon N, Poolsup N, Boonkaew S, Suthisisang CC. Meta-analysis of the effect of herbal supplement on glycemic control in type 2 diabetes. J Ethnopharmacol 2011; 137(3): 1328-33. [http://dx.doi.org/10.1016/j.jep.2011.07.059] [PMID: 21843614]

[40]   Kranti M, Mahesh V, Srinivas P, Ganesh Y, Godwin A, Lahkar M. Evaluation the protective effect of silymarin on doxorubicin induced chronic testicular toxicity in rats. Int J Pharma Bio Sci 2013; 4: 473-84.

[41]   Ramadan S, Shalaby M, Afifi N, El-Banna H. Hepatoprotective and antioxidant effects of *Silybum marianum* plant in rats. International Journal for Agro Veterinary and Medical 2011; 5: 541-7. [http://dx.doi.org/10.5455/ijavms.129]

[42]   Arbid MSS, Marquardt RR. Effects of intraperitoneally injected vicine and convicine on the rat: Induction of favism-like signs. J Sci Food Agric 1986; 37(6): 539-47. [http://dx.doi.org/10.1002/jsfa.2740370606]

[43]   Pan L, Fu JH, Xue XD, Xu W, Zhou P, Wei B. Melatonin protects against oxidative damage in a neonatal rat model of bronchopulmonary dysplasia. World J Pediatr 2009; 5(3): 216-21. [http://dx.doi.org/10.1007/s12519-009-0041-2] [PMID: 19693467]

[44]   El-Adawi H, El-Azhary D, Abd El-Wahab A, El-Shafeey M, Abdel-Mohsen M. Protective effect of milk thistle and grape seed extracts on fumonisin B1 induced hepato- and nephrotoxicity in rats. Egyptian Academic Journal of Biological Sciences 2011; 5: 6316-27. [http://dx.doi.org/10.5897/JMPR11.927]

[45]   Abdalla AA, Abou-Shehema BM, Hamed RS, El-Deken MR. Effect of silymarin supplementation on the performance of developed chickens under summer conditions. 1 – during growth period. Egypt Poult Sci 2018; 38: 305-29. [http://dx.doi.org/10.21608/epsj.2018.5667]

[46]   Khazaei R, Seidavi A, Bouyeh M. A review on the mechanisms of the effect of silymarin in milk thistle (*Silybum marianum*) on some laboratory animals. Vet Med Sci 2022; 8(1): 289-301. [http://dx.doi.org/10.1002/vms3.641] [PMID: 34599793]

[47]   Sabiu S, Sunmonu TO, Ajani EO, Ajiboye TO, Ajiboyeb O. Combined administration of silymarin and vitamin C stalls acetaminophen-mediated hepatic oxidative insults in Wistar rats. Rev Bras Farmacogn 2015; 25(1): 29-34.
[http://dx.doi.org/10.1016/j.bjp.2014.11.012]

[48]   Gobalakrishnan S, Asirvatham SS, Janarthanam V. Effect of silybin on lipid profile in hypercholesterolaemic rats. J Clin Diagn Res 2016; 10(4): FF01-5.
[http://dx.doi.org/10.7860/JCDR/2016/16393.7566] [PMID: 27190826]

[49]   Škottová N, Večeřa R, Urbánek K, Váňa P, Walterová D, Cvak L. Effects of polyphenolic fraction of silymarin on lipoprotein profile in rats fed cholesterol-rich diets. Pharmacol Res 2003; 47(1): 17-26.
[http://dx.doi.org/10.1016/S1043-6618(02)00252-9] [PMID: 12526857]

[50]   Phillips CJC, Hosseintabar-Ghasemabad B, Gorlov IF, Slozhenkina MI, Mosolov AA, Seidavi A. Immunomodulatory effects of natural feed additives for meat chickens. Life (Basel) 2023; 13(6): 1287.
[http://dx.doi.org/10.3390/life13061287] [PMID: 37374069]

[51]   Tedesco DEA, Guerrini A. Use of milk thistle in farm and companion animals: a review. Planta Med 2023; 89(6): 584-607.
[http://dx.doi.org/10.1055/a-1969-2440] [PMID: 36302565]

[52]   Bendich A. Physiological role of antioxidants in the immune system. J Dairy Sci 1993; 76(9): 2789-94.
[http://dx.doi.org/10.3168/jds.S0022-0302(93)77617-1] [PMID: 8227682]

[53]   Chand N, Muhammad D, Durrani F, Subhan Q, Ullah S. Protective effects of milk thistle (*Silybum marianum*) against aflatoxin B1 in broiler chicks. Asian-Aust. J Anim Sci 2011; 24: 1011-8.
[http://dx.doi.org/10.5713/ajas.2011.10418]

[54]   Makki O, Afzali N, Omidi A. Effect of milk thistle on the immune system, intestinal related variables, appearance and mortality of broilers contaminated with aflatoxin B1. J Herb Drugs 2013; 4: 33-8. Available from: jhd.shahrekord.iau.ir/article_633119_edac27add0eae669effc296adbecd586.pdf].

[55]   Tedesco D, Steidler S, Galletti S, Tameni M, Sonzogni O, Ravarotto L. Efficacy of silymarin-phospholipid complex in reducing the toxicity of aflatoxin B1 in broiler chicks. Poult Sci 2004; 83(11): 1839-43.
[http://dx.doi.org/10.1093/ps/83.11.1839] [PMID: 15554059]

[56]   Kalorey DR, Kurkure NV, Ramgaonkar JS, Sakhare PS, Warke S, Nigot NK. Effect of polyherbal feed supplement "Growell" during induced aflatoxicosis, ochratoxicosis and combined mycotoxicoses in broilers. Asian-Australas J Anim Sci 2005; 18(3): 375-83.
[http://dx.doi.org/10.5713/ajas.2005.375]

[57]   Basaga H, Poli G, Tekkaya C, Aras I. Free radical scavenging and antioxidative properties of 'silibin' complexes on microsomal lipid peroxidation. Cell Biochem Funct 1997; 15(1): 27-33.
[http://dx.doi.org/10.1002/(SICI)1099-0844(199703)15:1<27::AID-CBF714>3.0.CO;2-W] [PMID: 9075334]

[58]   Mueller A, Wolfe H, Meyer R. Precipitin production in bursectomised chickens and chickens injected with 19-nortestosterone on the fifth d of incubation. J Immunol 1960; 85: 172-9.
[http://dx.doi.org/10.4049/jimmunol.85.2.172] [PMID: 14424904]

[59]   Attia YA, Hamed RS, Bovera F, Abd El-Hamid AEHE, Al-Harthi MA, Shahba HA. Semen quality, antioxidant status and reproductive performance of rabbits bucks fed milk thistle seeds and rosemary leaves. Anim Reprod Sci 2017; 184: 178-86.
[http://dx.doi.org/10.1016/j.anireprosci.2017.07.014] [PMID: 28765034]

[60]   Collodel G, Moretti E, Noto D, Corsaro R, Signorini C. Oxidation of polyunsaturated fatty acids as a promising área of research in infertility. Antioxidants 2022; 11(5): 1002.
[http://dx.doi.org/10.3390/antiox11051002] [PMID: 35624866]

[61]   Aitken RJ. Reactive oxygen species as mediators of sperm capacitation and pathological damage. Mol

Reprod Dev 2017; 84(10): 1039-52.
[http://dx.doi.org/10.1002/mrd.22871] [PMID: 28749007]

[62] Lee SH, Lee S. Effects of melatonin and silymarin on reactive oxygen species, nitric oxide production, and sperm viability and motility during sperm freezing in pigs. Animals (Basel) 2023; 13(10): 1705.
[http://dx.doi.org/10.3390/ani13101705] [PMID: 37238134]

[63] Oufi H, Al-Shawi N, Hussain S. What are the effects of silibinin on testicular tissue of mice? J Appl Pharm Sci 2012; 2: 9-13. [http://dx.doi.org/10.7324/JAPS.2012.21103].

[64] Khalil EAM. Hormonal profile and histopathological study on the influence of silymarin on both female and male albino rats. Egypt J Hosp Med 2003; 13(1): 112-22.
[http://dx.doi.org/10.21608/ejhm.2003.18236]

[65] Muthu M, Prabu SM. Silibinin ameliorates oxidative stress mediated testicular damage by arsenic in rats. Asian Pac J Trop Biomed 2012; 1: 1-7.

[66] Hamed RS, Attia YA, Abd El-Hamid EA, Shahba HA. Impact of supplementation with milk thistle seeds and rosemary leaves on semen quality, antioxidant status and reproductive performance of rabbit bucks. Egyptian Poultry Science Journal 2016; 36: 279-98.
[http://dx.doi.org/10.21608/epsj.2016.33268]

[67] Abid Ali WD, Khudair ARN, Al-Masoudi EA. Ameliorative role of silymarin extracted from *Silybum marinum* seeds on nickel chloride induce changes in testicular functions in adult male rabbits. J Vet Res (Pulawy) 2015; 14: 135-44. Available from: [www.iasj.net/iasj/download/30bef9353f2130f1].

[68] Farmer C, Lapointe J, Palin MF. Effects of the plant extract silymarin on prolactin concentrations, mammary gland development, and oxidative stress in gestating gilts. J Anim Sci 2014; 92(7): 2922-30.
[http://dx.doi.org/10.2527/jas.2013-7118] [PMID: 24504042]

[69] Refaie AM, Ghazal MN, Abo El-Azayem EH, Abd El-maged MH. Impact of dietary supplementation of milk thistle (*Silybum marianum*) seed extract on doe rabbits performance. Egypt J Nutr Feeds 2019; 22(2): 375-82.
[http://dx.doi.org/10.21608/ejnf.2019.79433]

[70] Capasso R, Aviello G, Capasso F, *et al.* Silymarin BIO-C®, an extract from *Silybum marianum* fruits, induces hyperprolactinemia in intact female rats. Phytomedicine 2009; 16(9): 839-44.
[http://dx.doi.org/10.1016/j.phymed.2009.02.007] [PMID: 19303749]

[71] Mohammad B, Alzamely H, Gharrawi FA, Al-Aubaidy HAM. Milk thistle seed extract favorably affects lactation and development of mammary gland in female ras. Egypt J Vet Sci 2019; 50: 27-36.
[http://dx.doi.org/10.21608/ejvs.2018.6754.1058]

[72] Jawad DH, Albazi WJ, Mahmood HB. Role of silymarin on the some serum sex hormones and oxidantantioxidant parameters in during lactation period on female rats. Int J Health Sci 2022; 6: 5262-9.
[http://dx.doi.org/10.53730/ijhs.v6nS5.9778]

[73] Ben-Jonathan N, Hnasko R. Dopamine as a prolactin (PRL) inhibitor. Endocr Rev 2001; 22(6): 724-63.
[http://dx.doi.org/10.1210/edrv.22.6.0451] [PMID: 11739329]

[74] Gragnoli C, Reeves GM, Reazer J, Postolache TT. Dopamine–prolactin pathway potentially contributes to the schizophrenia and type 2 diabetes comorbidity. Transl Psychiatry 2016; 6(4): e785.
[http://dx.doi.org/10.1038/tp.2016.50] [PMID: 27093067]

[75] Loisel F, Quesnel H, Farmer C. Short Communication: Effect of silymarin (*Silybum marianum*) treatment on prolactin concentrations in cyclic sows. Can J Anim Sci 2013; 93(2): 227-30.
[http://dx.doi.org/10.4141/cjas2012-024]

[76] Nachammai V, Jeyabalan S, Muthusamy S. Anxiolytic effects of silibinin and naringenin on zebrafish model. Indian J Pharmacol 2021; 53(6): 457-64.
[http://dx.doi.org/10.4103/ijp.IJP_18_20] [PMID: 34975133]

[77]   Di Pierro F, Callegari A, Carotenuto D, Tapia MM. Clinical efficacy, safety and tolerability of BIO-C (micronized Silymarin) as a galactagogue. Acta Biomed 2008; 79(3): 205-10. Available from: mattioli1885journals.com/index.php/actabiomedica/article/view/1249/894].
[PMID: 19260380]

[78]   Gori L, Gallo E, Mascherini V, Mugelli A, Vannacci A, Firenzuoli F. Can estragole in fennel seed decoctions really be considered a danger for human health? A fennel safety update. Evid Based Complement Alternat Med 2012; 2012: 1-10.
[http://dx.doi.org/10.1155/2012/860542] [PMID: 22899959]

[79]   Tedesco D, Tameni M, Steidler S, Galletti S, Di Pierro F. Effect of silymarin and its phospholipid complex against AFM1 excretion in an organic dairy herd. Milchwissenschaft 2003; 58: 416-9.

[80]   Tedesco D, Domeneghini C, Sciannimanico D, Tameni M, Steidler S, Galletti S. Silymarin, a possible hepatoprotector in dairy cows: biochemical and histological observations. J Vet Med A Physiol Pathol Clin Med 2004; 51(2): 85-9.
[http://dx.doi.org/10.1111/j.1439-0442.2004.00603.x] [PMID: 15153078]

[81]   Tedesco D, Tava A, Galletti S, *et al.* Effects of silymarin, a natural hepatoprotector, in periparturient dairy cows. J Dairy Sci 2004; 87(7): 2239-47.
[http://dx.doi.org/10.3168/jds.S0022-0302(04)70044-2] [PMID: 15328238]

[82]   Vojtíšek B, Hronová B, Hamřík J, Janková B. Milk thistle (*Silybum marianum*, L.) in feed rations administered to ketotic cows. Vet Med (Praha) 1991; 36: 321-30.
[PMID: 1807027]

[83]   Křižová L, Watzková1 J, Třináctý J, Richter M, Buchta M. Rumen degradability and whole tract digestibility of flavonolignans from milk thistle (*Silybum marianum*) fruit expeller in dairy cows. Czech Journal of Animal Science 2011; 56: 269-278. Available from: https://www.agriculture journals.cz/pdfs/cjs/2011/06/03.pdf

CHAPTER 7

Turmeric (*Curcuma longa*)

Youssef A. Attia[1,2]**, Mohamed E. Abd El-Hack**[3,*]**, Mahmoud M. Alagawany**[3]**, Adel D. Al-qurashi**[2]**, Khalid A. Asiry**[2]**, Abdulmohsen H. Alqhtani**[4]**, Bahaa Abou-Shehema**[5]**, Ahmed A. Abdallah**[5]**, Ayman E. Taha**[6] **and Vincenzo Tufarelli**[7]

[1] *Animal and Poultry Production Department, Faculty of Agriculture, Damanhour University, Damanhour-22713, Egypt*

[2] *Sustainable Agriculture Production Research Group, Agriculture Department, Faculty of Environmental Sciences, King Abdulaziz University, Jeddah-21589, Saudi Arabia*

[3] *Poultry Department, Faculty of Agriculture, Zagazig University, Zagazig-44519, Egypt*

[4] *Department of Animal Production, College of Food and Agricultural Sciences, King Saud University, Riyadh, Saudi Arabia*

[5] *Department of Poultry Nutrition, Department of Rabbits and Waterfowl Breeding, Animal Production Institute, Agricultural Research Center, Dokki, Giza-3751310, Egypt*

[6] *Department of Animal Husbandry and Animal Wealth Development, Faculty of Veterinary Medicine, Alexandria University, Apis, Alexandria, 21944, Egypt*

[7] *Department of Precision and Regenerative Medicine and Jonian Area (DiMePRe-J), Section of Veterinary Science and Animal Production, University of Bari Aldo Moro, 70010 Valenzano, Bari, Italy*

Abstract: Two molecules of feruloyl-CoA and one molecule of malonyl-CoA are combined to generate turmeric (*Curcuma longa*), also known as curcumin, through two enzymatic processes mediated by curcumin synthase (CURS) and DIKETIDE-CoA SYNTHASE (DCS). DCS and CURS are members of polyketide synthase family type III. Turmeric, a homegrown spice, has several health benefits in the medical field. One specific bioactive ingredient produced by turmeric is curcumin, a polyphenolic phytochemical with antibacterial, anti-inflammatory, anticancer, and antioxidant properties. Research indicates that turmeric can substitute for antibiotics in chicken feed and is effective. When powdered turmeric rhizomes are fed to broiler chicks, morbidity and mortality are reduced. Furthermore, it has been shown that including turmeric in chicken feed does not negatively impact the overall health of animals. The use of turmeric in animal nutrition as a helpful feed additive, as well as its bioactive components and effects on blood biochemistry, animal health, and productive performance as an antibiotic substitute, will be covered in this chapter.

---

* Corresponding author Mohamed E. Abd El-Hack: Poultry Department, Faculty of Agriculture, Zagazig University, Zagazig-44519, Egypt; E-mail: dr.mohamed.e.abdalhaq@gmail.com

Keywords: Animal health, Bioactive substances, Growth promoter, Livestock, Productive performance, Turmeric.

INTRODUCTION

Over the past 10 years, commercially available antibiotics have been added to chicken feed to aid in defense against exogenous infections [1 - 3]. Antibiotics may be harmful to public health even if they help reduce morbidity and mortality issues related to the production of poultry [4, 5]. Since January 2006, antibiotics have been prohibited in poultry feed across Europe [4]. As a result, to produce chicken feed without antibiotics, enterprises must find necessary alternatives [6]. Antibiotic-free meals are supplemented with a variety of bioactive substances to enhance animal production and quality of life [7 - 9].

Medical biology makes extensive use of the household spice turmeric (*Curcuma longa*); (Fig. **1**) [10]. One specific bioactive ingredient of turmeric is curcumin, a polyphenolic phytochemical with antibacterial, anti-inflammatory, anticancer, and antioxidant activities [10 - 12]. Present research indicates that turmeric may replace antibiotics in chicken feed and is successful in doing so [1]. The addition of turmeric rhizome powder to poultry feed lowers the morbidity and mortality of meat-type chickens [10]. Furthermore, it has been shown that turmeric in chicken feed does not negatively affect the health of animals [1].

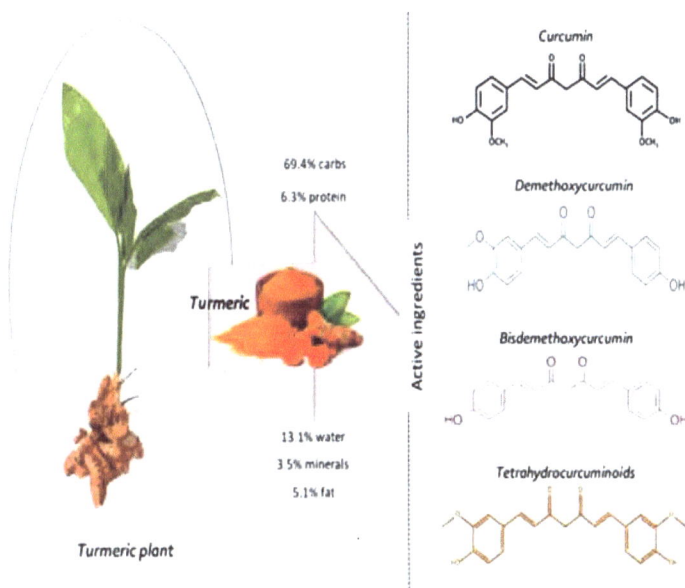

Fig. (1). The main bioactive ingredients present in turmeric extract.

Two molecules of feruloyl-CoA and one molecule of malonyl-CoA combine to generate curcumin through two enzymatic processes mediated by DIKETIDE-CoA SYNTHASE (DCS) and CURCUMIN SYNTHASE (CURS). The type III polyketide synthase family includes DCS and CURS [13 - 15].

Importance of Turmeric Feed in Poultry Nutrition

Humans employ an abundant supply of bioactive compounds found in turmeric rhizomes for both medicinal and non-medicinal purposes [16, 17]. Turmeric is a natural, safe, and appropriate dietary supplement that is often included in the daily diet, as opposed to antibiotics that are easily obtained on the market [17, 18]. Turmeric is composed of 69.4% carbohydrates, 6.3% protein, 5.1% fat, 3.5% minerals, and 13.1% water [19, 20]. It also contains large amounts of the phenolic constituents curcumin, dimethoxy curcumin, bisdemethoxy curcumin, and metabolites of tetrahydrocurcumin [21, 22]. Turmeric polyphenolic molecules have a wide range of biological activities, including antiviral, anti-inflammatory, antibacterial, antihypertensive, anti-carcinogenic, and anti-fungal activities [23, 24].

The body weight and growth rate of broiler chicks were greatly boosted by adding turmeric powder to their food, according to the literature that is now accessible [26]. Turmeric supplements stimulate the digestive system by increasing intestinal maltase, lipase, and sucrose activity as well as pancreatic lipase, amylase, trypsin, and chymotrypsin synthesis [27, 28]. Evidence suggests the beneficial effects of turmeric on chicken egg production [29]. Supplementation of the diet with turmeric increased egg production, yolk weight, and yolk index [29]. However, the effectiveness of turmeric as a natural growth booster has been debated. This section reviews the literature that has been reported in this field.

Bio-active Compounds

Turmeric contains a variety of bioactive compounds, including curcumin, demethoxycurcumin, bisdemethoxycurcumin, and tetrahydrocurcuminoids [11, 14, 30 - 32]. These compounds have antioxidant, anti-inflammatory, and nematocidal properties [13, 15, 30, 33 - 37], as well as protective properties against aflatoxin-induced hepatocarcinogenesis, mutagenicity [38, 39], and coccidiosis [40 - 43]. (Fig. **1**) summarizes the bioactive components of the turmeric plants.

Growth Performance

The benefits of supplementing chickens with turmeric at 0–10 g/kg have been inconsistently reported. For instance, broilers fed diets containing 5 g/kg turmeric showed more productive features [11, 44, 45]. Furthermore, compared to the control group, Attia *et al.* [46] found that adding 1 g/kg turmeric to the diet significantly increased growth. It also enhanced the feed conversion ratio (FCR) and European production index but had no influence on the feed consumption of broiler groups when compared to the MOS and control groups. According to these authors, mannooligosaccharides or oxytetracycline may be substituted with turmeric at a rate of 1 kg/ton feed as a phytogenic feed addition without negatively impacting the economic and productive characteristics of broilers. However, turmeric powder had no discernible impact on the growth traits or carcass production of broiler chickens [31, 47, 48]. (Fig. **2**) highlights the significance of adding turmeric to the chicken diet.

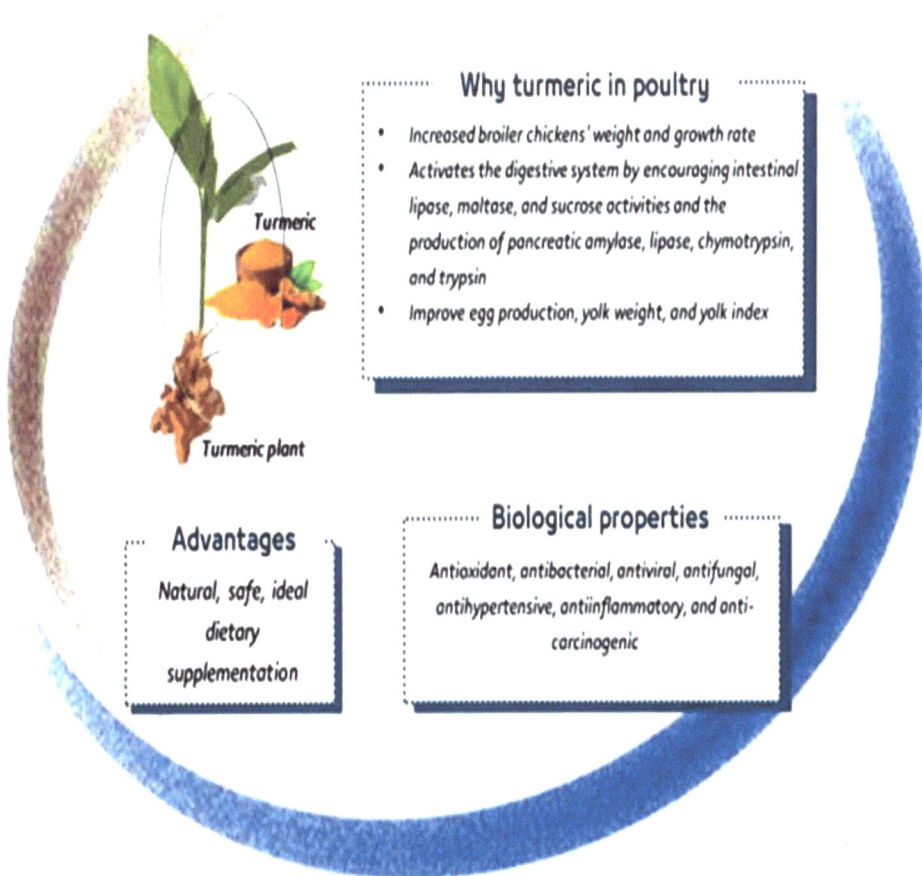

Why turmeric in poultry

- Increased broiler chickens' weight and growth rate
- Activates the digestive system by encouraging intestinal lipase, maltase, and sucrose activities and the production of pancreatic amylase, lipase, chymotrypsin, and trypsin
- Improve egg production, yolk weight, and yolk index

Turmeric

Turmeric plant

Advantages

Natural, safe, ideal dietary supplementation

Biological properties

Antioxidant, antibacterial, antiviral, antifungal, antihypertensive, antiinflammatory, and anti-carcinogenic

Fig. (2). Importance of turmeric supplementation in poultry diet.

Blood Constituents

The serum protein profiles and ALKP, ALT, and AST enzymes were unaffected by a 0–10 g/kg dosage of turmeric [11, 44, 45]. Turmeric powder at 0.6 and 0.9 g/kg, according to Ahmadi [39], enhanced antioxidant defense enzymes including superoxide dismutase and catalase and decreased MDA. It also reduced the detrimental effects of aflatoxin B1 on the serum total protein profiles. Broiler feed with 5 g/kg turmeric powder significantly lowered the levels of liver enzymes (ALT and ALKP) [45]. Supplementing chickens with turmeric had positive effects on the biochemical blood metabolites. For instance, 0.35 g/kg turmeric meal increased the formation of serum HDL, which in turn decreased the levels of total cholesterol, LDL, and VLDL [50].

Turmeric supplementation at 3.3, 6.6, and 10 g/kg feed considerably reduced plasma triglyceride levels compared to the control group, but had no discernible influence on total cholesterol, LDL-, and HDL-cholesterol, according to Nouzarian *et al.* [48]. When compared to the other groups, Attia *et al.* [46] observed that supplementation with turmeric at doses of 1 and 2 g/kg resulted in a substantial decrease in total protein. Additionally, compared to groups supplemented with mannooligosaccharides and oxytetracycline, the 1 g/kg group supplemented with turmeric showed a reduction in blood globulin. The group supplemented with various additives exhibited a substantial increase in alkaline phosphatase compared to the control group; the group administered 1 g/kg turmeric showed the largest impact. AST was considerably reduced by 0.5 g/kg of turmeric as compared to the oxytetracycline and control groups.

Furthermore, the AST/ALT ratio of the turmeric-fortified group was much lower than that of the oxytetracycline group. The TAC of the supplemented groups was notably greater than that of the control group; the group receiving turmeric (0.5 g) had the highest TAC. There were no appreciable differences in MAD across the groups. All blood hematological features were considerably affected by other therapies, with the exception of MCV and RBCs, according to the findings of the authors. Within the MOS group, the PCV and Hgb levels were the lowest, while those of turmeric at 1 g/kg were the highest. Moreover, MCH and MCHC were highest in the control group and lowest in the MOS group, respectively. The findings showed that the blood hematological parameters of the turmeric group at 1 g/kg were comparable to those of the control group. Owing to its antioxidative properties, turmeric also enhances the oxidative stability of broiler meat [51].

Lipolysis Effect

Rats treated with curcumin and an ether extract of turmeric showed hypolipemic effects [52, 53] as well as reductions in triglycerides, cholesterol, and fatty acids in alcohol-induced toxicity. *Curcuma domestica* reduced the levels of triglycerides in the blood of animals fed a high-cholesterol diet but had no impact on the levels of cholesterol or phospholipids [54]. The addition of turmeric to broiler chicken meals resulted in a considerable increase in both total and HDL cholesterol. At 42 days of age, it reduced LDL cholesterol, but had no effect on total protein, total triglycerides, or hematocrit values [49]. Liver fatty acid synthase activity, which may be affected by medicinal plants and result in altered lipid and lipoprotein metabolism, may be the reason for the decrease in lipid synthesis caused by turmeric powder [48]. Rats administered curcumin had significantly lower liver cholesterol and triacylglycerol levels than control rats fed a high-fat diet [55]. They also concluded that rats administered curcumin had decreased plasma triacylglycerol levels in the LDL-c fraction. The hepatic acyl-CoA oxidase activity in the curcumin group was noticeably greater than that in the control group.

Because livestock production farming is under pressure to improve animal reproductive and productive performance, increase economic profits, and improve the safety of animal products for human consumption, curcumin is increasingly ideal for animal feed as a substitute for chemical supplements, such as antibiotics and chemotherapeutic drugs in the animal industry [2, 32]. By regulating lipid metabolism in broilers, curcumin supplementation reduces the absolute and relative weights of belly fat [56]. Curcumin supplementation improves meat quality, reduces oxidative stress, and prevents pigs from gaining too much fat [57]. Curcumin has shown remarkable benefits in a variety of animal husbandry applications, such as reproductive and productive performance, digestibility and feed utilization, immune response, stress tolerance, and histopathology in various age groups of single-stomach animals, including fish, poultry, and pigs [58].

Immune Response

Natural plant-based solutions are interesting alternatives to synthetic chemical insecticides for pest control, and can be utilized as growth enhancers in cattle and poultry performance in place of antibiotics [59]. The natural polyphenol component, curcumin, is derived from the rhizomes of turmeric (*Curcuma* spp.). It qualifies for its insecticidal and nutritional properties and offers a number of medicinal advantages in the treatment of human ailments [59].

It has been discovered that curcumin functions as an antioxidant and an inflammatory agent [13 - 15, 37]. However, supplementing hens with turmeric powder may not be the best way to boost their immune system [48]. Curcumin at doses of 10 and 20 mg/kg had no effect on NK cell IgG levels. On the other hand, whereas 10, 20, and 40 mg/kg did not cause delay-type hypersensitivity or NK cell activity, a larger dose (40 mg/kg) considerably increased IgG levels [60]. These variations in the manner in which various batches of turmeric powder respond to supplements might be the cause of these discrepancies. To this end, four further collections of turmeric showed a range of 1.06 to 5.70% for curcumin [61]. The contrasting and irregular immunological responses to turmeric observed in different studies might be attributed to the health condition of the birds, the cleanliness of the experimental site, outside obstacles, and the digestibility and composition of their basal food [62].

Aflatoxin B1's detrimental effects on the immune system have been studied in relation to turmeric supplementation. A study that examined the effects of turmeric at 0.5 g/kg feed on the immune system found that turmeric has the capacity to stimulate the humoral immune system. Turmeric also reduced the detrimental effects of ochratoxin A on the immunological and hematobiochemical systems of laying hens [64]. Birds fed diets containing turmeric showed significantly higher serum antibody levels against antiamicroneme protein 2 from Eimeria tenella (EtMIC2), suggesting that turmeric has immunomodulatory activities against parasite infections [41, 65]. An apical complex protein, EtMIC2, is necessary for Eimeria parasites to invade host cells. It is essential for preventing sporozoite penetration by host cells and may be useful in preventing parasite adherence to the cell. According to Yarru *et al.* [66], turmeric at 5.0 g/kg feed increased the expression of genes involved in immunological function (IL-6 and IL-2) and antioxidant activity (CYP450). The HI titer levels of the ND vaccine increased when broilers were fed a diet enriched with turmeric [67]. According to HA and HI tests, feeding broilers a diet enriched with 1 g/kg of *Curcuma longa* increased the humoral response to RD vaccination [26]. When broilers were given powdered turmeric rhizomes, there was a decrease in the ratio of monocytes and an increase in IgA, IgM, and IgG [68]. Furthermore, feeding broiler turmeric was linked to a decrease in HI titer levels against Newcastle disease (NDV) [47]. Furthermore, according to Sadeghi *et al.*, compared to the control, turmeric infusion did not increase the antibody titer to NDV [31].

According to Lee *et al.* [69], when given dietary form, a combination of capsicum and turmeric oleoresins was shown to be a beneficial phytonutrient counter to clinical indications of avian necrotic enteritis. To further reduce necrotic enteritis, more research is needed to clarify the molecular and cellular properties of this phytochemical mixture.

The pharmacological characteristics of curcumin include wound healing and antiviral, antifungal, anti-inflammatory, and antioxidant effects [70]. Turmeric powder is also a strong immunomodulatory agent that can reduce the activation of dendritic cells, B cells, T cells, neutrophils, macrophages, and natural killer cells

[71]. Therefore, better immune responses and increased generation of antibody titers are anticipated [71]. However, no discernible trend was seen as a result of supplementation with turmeric between 3.3 and 10 g/kg diet, and treatments were unable to significantly affect the formation of antibody titers [48].

However, Emadi and Kermanshahi [68] discovered that adding turmeric powder to meals also had an impact on the serum immunoglobulins of hens, resulting in a substantial rise in IgA and IgM at 21 days of age as well as IgG at 21 and 42 days of age.

Supplementing with turmeric at doses of 5 and 10 g/kg increased leukocyte concentrations [11], whereas the immunostimulatory action of curcumin boosted WBC concentrations [72]. The active compounds found in turmeric stimulate bile flow and secretion, which supports liver function [68]. Additionally, studies have shown that curcumin increases the levels of the antioxidant enzymes catalase, superoxide dismutase, and glutathione peroxidase [66]. Turmeric powder at 0.6 and 0.9%, according to Ahmadi [39], increased antioxidant defense (catalase, superoxide dismutase, and reduced MDA) and mitigated the detrimental effects of AFB1 on blood total protein, albumin, and globulin.

Curcumin, the most effective chemical compound and ecological insecticide, is a bioactive compound with insecticidal activities found in turmeric essential oil [73]. The larvicidal properties of curcumin make it an environmentally benign agent [74].

Curcumin improves productivity and manages insect pests during cattle production According to recent research, curcumin is an effective feed ingredient that supports animal development and disease resistance in poultry and livestock [20, 51, 59]. Additionally, curcumin may have insecticidal and growth-inhibiting properties that reduce the spread of agricultural insect pests and human illnesses by insects [59, 73].

CONCLUSION AND REMARKS

The bioactive compounds in turmeric, particularly curcumin, exhibit antioxidant, anti-inflammatory, and immunomodulatory properties that can positively affect growth performance, blood constituents, and immune responses in poultry. Although the results have been inconsistent across studies, turmeric supplementation has generally shown improvements in productive performance, feed conversion ratio, and certain blood parameters. Its effects on lipid metabolism and immune function are particularly noteworthy, although further research is required to fully elucidate these mechanisms. The use of turmeric in poultry nutrition represents a sustainable approach to enhancing animal health and productivity without relying on antibiotics. However, standardization of dosage and further investigation of optimal supplementation strategies are necessary to maximize the benefits of turmeric in poultry feed. As the livestock industry continues to seek alternatives to antibiotics, turmeric stands out as a promising natural feed additive, worthy of continued research and application.

IMPLICATIONS

- Turmeric could serve as a viable alternative to antibiotics in poultry feed, addressing concerns about antibiotic resistance, while potentially improving animal health and productivity.
- The bioactive compounds in turmeric, particularly curcumin, offer multiple benefits including antioxidant, anti-inflammatory, and immunomodulatory effects, which could lead to improved overall health and performance in poultry.
- Turmeric supplementation may enhance growth performance, feed conversion ratio, and certain blood parameters in poultry, although the results have been inconsistent across studies, indicating a need for further research to optimize dosage and application.
- The potential of turmeric to positively influence lipid metabolism in poultry could lead to improvements in meat quality and reduced fat deposition, which are desirable outcomes in the poultry industry.
- The immunomodulatory effects of turmeric suggest that it could enhance disease resistance in poultry, potentially reducing the need for medical interventions and improving the overall flock health.
- The use of turmeric as a natural feed additive aligns with the growing consumer demand for antibiotic-free and naturally raised poultry products, potentially offering market advantages to organic poultry producers.
- Further research is needed to standardize the dosages, understand the full mechanisms of action, and develop optimal supplementation strategies to maximize the benefits of turmeric in poultry nutrition.
- The multifaceted benefits of turmeric in poultry nutrition highlight the potential of phytogenic feed additives in sustainable livestock production, opening avenues for further exploration of plant-based solutions for animal husbandry.
- At 1–10 g/kg food, turmeric has been utilized as a natural growth enhancer; however, depending on the chicken strain and the desired performance, the results are still debatable.

REFERENCES

[1]   Dono ND. Turmeric (*Curcuma longa* Linn) supplementation as an alternative to antibiotics in poultry diets (2014). WARTAZOA. Indonesian Bulletin of Animal and Veterinary Sciences 2014; 23(1): 41-9. [http://dx.doi.org/10.14334/wartazoa.v23i1.958]

[2]   Shehata AA, Yalçın S, Latorre JD, *et al.* Probiotics, prebiotics, and phytogenic substances to optimize gut health in poultry. Microorganisms 2022; 10(2): 395.

[http://dx.doi.org/10.3390/microorganisms10020395] [PMID: 35208851]

[3]   Tan Z, Halter B, Liu D, Gilbert ER, Cline MA. Dietary flavonoids modulate lipid metabolism in poultry. Front Physiol 2022; 13: 863860.
[http://dx.doi.org/10.3389/fphys.2022.863860] [PMID: 35547590]

[4]   Casewell M, Friis C, Marco E, McMullin P, Phillips I. The European ban on growth-promoting antibiotics and emerging consequences for human and animal health. J Antimicrob Chemother 2003; 52(2): 159-61.
[http://dx.doi.org/10.1093/jac/dkg313] [PMID: 12837737]

[5]   Shehata AA, Attia Y, Khafaga AF, *et al.* Restoring healthy gut microbiome in poultry using alternative feed additives with particular attention to phytogenic substances: Challenges and prospects. German Journal of Veterinary Research 2022; 2(3): 32-42.
[http://dx.doi.org/10.51585/gjvr.2022.3.0047]

[6]   Çabuk M, Bozkurt M, Alçiçek A, Akbaþ Y, Küçükyýlmaz K. Effect of a herbal essential oil mixture on growth and internal organ weight of broilers from young and old breeder flocks. S Afr J Anim Sci 2006; 36(2): 135-41.
[http://dx.doi.org/10.4314/sajas.v36i2.3996]

[7]   Attia YA, Alagawany MM, Farag MR, *et al.* Phytogenic products and phytochemicals as a candidate strategy to improve tolerance to COVID-19. Front Vet Sci 2020; 7: 573159.
[http://dx.doi.org/10.3389/fvets.2020.573159]

[8]   Alagawany M, Attia YA, Farag Mayada R, *et al.* The strategy of boosting the immune system of food-producing during the CoViD-19 pandemic. Front Vet Sci 2020.
[http://dx.doi.org/10.3389/fvets.2020.570748]

[9]   He Y, Fu Z, Dai S, Yu G, Ma Z. Dietary curcumin supplementation can enhance health and resistance to ammonia stress in the greater amberjack (*Seriola dumerili*). Front Mar Sci 2022; 9: 961783.
[http://dx.doi.org/10.3389/fmars.2022.961783]

[10] Aggarwal BB, Harikumar KB. Potential therapeutic effects of curcumin, the anti-inflammatory agent, against neurodegenerative, cardiovascular, pulmonary, metabolic, autoimmune and neoplastic diseases. Int J Biochem Cell Biol 2009; 41(1): 40-59.
[http://dx.doi.org/10.1016/j.biocel.2008.06.010] [PMID: 18662800]

[11] Al-Sultan SI. The effect of *Curcuma longa* (turmeric) on overall performance of broiler chickens. Int J Poult Sci 2003; 2(5): 351-3.
[http://dx.doi.org/10.3923/ijps.2003.351.353]

[12] Abd El-Hack ME, El-Saadony MT, Swelum AA, *et al.* Curcumin, the active substance of turmeric: its effects on health and ways to improve its bioavailability. J Sci Food Agric 2021; 101(14): 5747-62.
[http://dx.doi.org/10.1002/jsfa.11372] [PMID: 34143894]

[13] Oyarce P, De Meester B, Fonseca F, *et al.* Introducing curcumin biosynthesis in *Arabidopsis* enhances lignocellulosic biomass processing. Nat Plants 2019; 5(2): 225-37.
[http://dx.doi.org/10.1038/s41477-018-0350-3] [PMID: 30692678]

[14] Wu J, Chen W, Zhang Y, Zhang X, Jin JM, Tang SY. Metabolic engineering for improved curcumin biosynthesis in *Escherichia coli.* J Agric Food Chem 2020; 68(39): 10772-9.
[http://dx.doi.org/10.1021/acs.jafc.0c04276] [PMID: 32864959]

[15] De Meester B, Oyarce P, Vanholme R, *et al.* Engineering curcumin biosynthesis in poplars affects lignification and biomass yield. Front Plant Sci 2022; 13: 943349.
[http://dx.doi.org/10.3389/fpls.2022.943349] [PMID: 35860528]

[16] Al-Kassie GAM, Mohseen AM, Abd Al-Jaleel RA. Modification of productive performance and physiological aspects of broilers by the addition of a mixture of cumin and turmeric to the diet. Res Opin Anim Vet Sci 2011; 1: 31-4.

[17] Khan RU, Naz S, Javdani M, *et al.* The use of turmeric (*Curcuma longa*) in poultry feed. Worlds Poult

Sci J 2012; 68(1): 97-103.
[http://dx.doi.org/10.1017/S0043933912000104]

[18]   Wang R, Li D, Bourne S. Can years of herbal medicine history help us solve problems in the year 2000. In: Lyons TP, Jacques KA, Eds. Proceedings of Alltech's 14th Annual Symposium. 273-91.

[19]   Chattopadhyay I, Biswas K, Bandyopadhyay U, Banerjee RK. Turmeric and curcumin: Biological actions and medicinal applications. Curr Sci 2004; 87(1): 44-53.

[20]   Zhang HA, Kitts DD. Turmeric and its bioactive constituents trigger cell signaling mechanisms that protect against diabetes and cardiovascular diseases. Mol Cell Biochem 2021; 476(10): 3785-814.
[http://dx.doi.org/10.1007/s11010-021-04201-6] [PMID: 34106380]

[21]   Roughley PJ, Whiting DA. Curcumin biosynthesis experiments J Chem Soc Perkin Trans 1973; 20: 2379-2388.

[22]   Huang MT, Ma W, Lu YP, *et al.* Effects of curcumin, demethoxycurcumin, bisdemethoxycurcumin and tetrahydrocurcumin on 12- *O* -tetradecanoylphorbol-13-acetateinduced tumor promotion. Carcinogenesis 1995; 16(10): 2493-7.
[http://dx.doi.org/10.1093/carcin/16.10.2493] [PMID: 7586157]

[23]   Masuda T, Maekawa T, Hidaka K, Bando H, Takeda Y, Yamaguchi H. Chemical studies on antioxidant mechanism of curcumin: analysis of oxidative coupling products from curcumin and linoleate. J Agric Food Chem 2001; 49(5): 2539-47.
[http://dx.doi.org/10.1021/jf001442x] [PMID: 11368633]

[24]   Rehman U, Syed QA, Asghar HA, Arshad MK, Sultan G, Asghar A, Aslam M, Abdullah M. Alimentary and Recuperative Prospective of *Curcuma longa* (Turmeric). Sch Int J Biochem. 2022;5(5):67-75.

[25]   Bhavani Shankar TN, Sreenivasa Murthy V. Effect of turmeric (*Curcuma longa*) fractions on the growth of some intestinal & pathogenic bacteria *in vitro*. Indian J Exp Biol 1979; 17(12): 1363-6.
[PMID: 540987]

[26]   Kumari P, Gupta MK, Ranjan R, Singh KK, Yadava R. *Curcuma longa* as feed additive in broiler birds and its patho-physiological effects. Indian J Exp Biol 2007; 45(3): 272-7.
[PMID: 17373373]

[27]   Platel K, Srinivasan K. Influence of dietary spices or their active principles on digestive enzymes of small intestinal mucosa in rats. Int J Food Sci Nutr 1996; 47(1): 55-9.
[http://dx.doi.org/10.3109/09637489609028561] [PMID: 8616674]

[28]   Platel K, Srinivasan K. Influence of dietary spices and their active principles on pancreatic digestive enzymes in albino rats. Food/ Nahrung. 2000; 44(1):42-46.
[http://dx.doi.org/10.1002/(SICI)1521-3803(20000101)44:1<42::AID-FOOD42>3.0.CO;2-D]

[29]   Radwan N, Hassan RA, Qota EM, Fayek HM. Effect of natural antioxidant on oxidative stability of eggs and productive and reproductive performance of laying hens. Int J Poult Sci 2008; 7(2): 134-50.
[http://dx.doi.org/10.3923/ijps.2008.134.150]

[30]   Kiuchi F, Goto Y, Sugimoto N, Akao N, Kondo K, Tsuda Y. Nematocidal activity of turmeric: synergistic action of curcuminoids. Chem Pharm Bull (Tokyo) 1993; 41(9): 1640-3.
[http://dx.doi.org/10.1248/cpb.41.1640] [PMID: 8221978]

[31]   Sadeghi GH, Karimi A, Padidar Jahromi SH, Azizi T, Daneshmand A. Effects of cinnamon, thyme and turmeric infusions on the performance and immune response in of 1- to 21-day-old male broilers. Rev Bras Cienc Avic 2012; 14(1): 15-20.
[http://dx.doi.org/10.1590/S1516-635X2012000100003]

[32]   Placha I, Gai F, Pogány Simonová M. Editorial: Natural feed additives in animal nutrition—Their potential as functional feed. Front Vet Sci 2022; 9: 1062724.
[http://dx.doi.org/10.3389/fvets.2022.1062724] [PMID: 36439337]

[33] Ammon HPT, Safayhi H, Mack T, Sabieraj J. Mechanism of antiinflammatory actions of curcumine and boswellic acids. J Ethnopharmacol 1993; 38(2-3): 105-12.
[http://dx.doi.org/10.1016/0378-8741(93)90005-P] [PMID: 8510458]

[34] Osawa T, Sugiyama Y, Inayoshi M, Kawakishi S. Antioxidative activity of tetrahydrocurcuminoids. Biosci Biotechnol Biochem 1995; 59(9): 1609-12.
[http://dx.doi.org/10.1271/bbb.59.1609] [PMID: 8520105]

[35] Wuthi-Udomler M, Grisanapan W, Luanratana O, Caichompoo W. Anti-fungal activities of plant extracts. Southeast Asian J Trop Med Public Health 2000; 31: 178-82.

[36] Alia BH, Marrif H, Noureldayemc SA, Bakheitd AO, Blunden G. Biological properties of curcumin: a review. NPC 2006; 1: 509-21.

[37] Karthikeyan A, Young KN, Moniruzzaman M, *et al.* Curcumin and its modified formulations on inflammatory bowel disease (Ibd): the story so far and future outlook. Pharmaceutics 2021; 13(4): 484.
[http://dx.doi.org/10.3390/pharmaceutics13040484] [PMID: 33918207]

[38] Soni K, Lahiri M, Chackradeo P, Bhide SV, Kuttan R. Protective effect of food additives on aflatoxin-induced mutagenicity and hepatocarcinogenicity. Cancer Lett 1997; 115(2): 129-33.
[http://dx.doi.org/10.1016/S0304-3835(97)04710-1] [PMID: 9149115]

[39] Ahmadi F. Effect of turmeric *Curcumin longa* powder on performance, oxidative stress state and some of blood parameters in broilers fed on diets containing aflatoxin. Glob Vet 2010; 5: 312-7.

[40] Allen PC, Fetterer RH. Recent advances in biology and immunobiology of Eimeria species and in diagnosis and control of infection with these coccidian parasites of poultry. Clin Microbiol Rev 2002; 15(1): 58-65.
[http://dx.doi.org/10.1128/CMR.15.1.58-65.2002] [PMID: 11781266]

[41] Lee K, Lillehoj HS, Siragusa GR. Direct-fed microbials and their impact on the intestinal microflora and immune system of chickens. J Poult Sci 2010; 47(2): 106-14.
[http://dx.doi.org/10.2141/jpsa.009096]

[42] Lee SH, Lillehoj HS, Jang SI, Kim DK, Ionescu C, Bravo D. Effect of dietary curcuma, capsicum, and lentinus on enhancing local immunity against *Eimeria acervulina* infection. J Poult Sci 2010; 47(1): 89-95.
[http://dx.doi.org/10.2141/jpsa.009025]

[43] Eevuri T, Putturu R. Use of certain herbal preparations in broiler feeds - A review. Vet World 2013; 6: 172-9.
[http://dx.doi.org/10.5455/vetworld.2013.172-179]

[44] Durrani FR, Ismail M, Sultan A, Suhail SM, Chand N, Durrani Z. Effect of different levels of feed added turmeric *Curcuma longa* on the performance of broiler chicks. J Agric Biol Sci 2006; 1: 9-11.

[45] Abou-Elkhair R, Ahmed HA, Selim S. Effects of black pepper (*piper nigrum*), turmeric powder (*curcuma longa*) and coriander seeds (*coriandrum sativum*) and their combinations as feed additives on growth performance, carcass traits, some blood parameters and humoral immune response of broiler chickens. Asian-Australas J Anim Sci 2014; 27(6): 847-54.
[http://dx.doi.org/10.5713/ajas.2013.13644] [PMID: 25050023]

[46] Attia YA, Al-Harthi MA, Hassan SS. Turmeric (*Curcuma longa* Linn.) as a phytogenic growth promoter alternative for antibiotic and comparable to mannan oligosaccharides for broiler chicks. Rev Mex Cienc Pecu 2017; 8(1): 11-21.
[http://dx.doi.org/10.22319/rmcp.v8i1.4309]

[47] Mehala C, Moorthy M. Effect of Aloe vera and *Curcuma longa* (Turmeric) on carcass characteristics and biochemical parameters of broilers. Int J Poult Sci 2008; 7(9): 857-61.
[http://dx.doi.org/10.3923/ijps.2008.857.861]

[48] Nouzarian R, Tabeidian S, Toghyani M, Ghalamkari G, Toghyani M. Effect of turmeric powder on

performance, carcass traits, humoral immune responses, and serum metabolites in broiler chickens. J Anim Feed Sci 2011; 20(3): 389-400.
[http://dx.doi.org/10.22358/jafs/66194/2011]

[49]   Emadi M, Kermanshahi H. Effect of turmeric rhizome powder on the activity of some blood enzymes in broiler chickens. Int J Poult Sci 2007; 6: 48-51.
[http://dx.doi.org/10.3923/ijps.2007.345.348]

[50]   Zhongze H, Like W, Aiyou W. Effect of curcumin on fat metabolism in Wanjiang Yellow chickens. Cerl. Feed Ind 2008; 4: 12.

[51]   Gharejanloo M, Mehri M, Shirmohammad F. Effect of different levels of turmeric and rosemary essential oils on performance and oxidative stability of broiler meat. Iran J Appl Anim Sci 2017; 7(4): 655-62.

[52]   Rao DS, Sekhara NC, Satyanarayana MN, Srinivasan M. Effect of curcumin on serum and liver cholesterol levels in the rat. J Nutr 1970; 100(11): 1307-15.
[http://dx.doi.org/10.1093/jn/100.11.1307] [PMID: 5476433]

[53]   Rukkumani R, Balasubashini MS, Menon VP. Protective effects of curcumin and photo-irradiated curcumin on circulatory lipids and lipid peroxidation products in alcohol and polyunsaturated fatty acid-induced toxicity. Phytother Res 2003; 17(8): 925-9.
[http://dx.doi.org/10.1002/ptr.1254] [PMID: 13680826]

[54]   Ahmad-Raus, Raha R, Elda-Surhaida ES, Abdul-Latif, Jamaludin J Mohammad. Lowering of lipid composition in aorta of guinea pigs by Curcuma domestica. BMC Complementary and Alternative Medicine 2001; 200, 1:6.

[55]   Asai A, Miyazawa T. Dietary curcuminoids prevent high-fat diet-induced lipid accumulation in rat liver and epididymal adipose tissue. J Nutr 2001; 131(11): 2932-5.
[http://dx.doi.org/10.1093/jn/131.11.2932] [PMID: 11694621]

[56]   Xie Z, Shen G, Wang Y, Wu C. Curcumin supplementation regulates lipid metabolism in broiler chickens. Poult Sci 2019; 98(1): 422-9.
[http://dx.doi.org/10.3382/ps/pey315] [PMID: 30053224]

[57]   Zhang J, Yan E, Zhang L, Wang T, Wang C. Curcumin reduces oxidative stress and fat deposition in *longissimus dorsi* muscle of intrauterine growth-retarded finishing pigs. Anim Sci J 2022; 93(1): e13741.
[http://dx.doi.org/10.1111/asj.13741] [PMID: 35707899]

[58]   Moniruzzaman M, Kim H, Shin H, *et al.* Evaluation of dietary curcumin nanospheres in a weaned piglet model. Antibiotics (Basel) 2021; 10(11): 1280.
[http://dx.doi.org/10.3390/antibiotics10111280] [PMID: 34827218]

[59]   Sureshbabu A, Smirnova E, Karthikeyan A, *et al.* The impact of curcumin on livestock and poultry animal's performance and management of insect pests. Front Vet Sci 2023; 10: 1048067.
[http://dx.doi.org/10.3389/fvets.2023.1048067] [PMID: 36816192]

[60]   South EH, Exon JH, Hendrix K. Dietary curcumin enhances antibody response in rats. Immunopharmacol Immunotoxicol 1997; 19(1): 105-19.
[http://dx.doi.org/10.3109/08923979709038536] [PMID: 9049662]

[61]   Jayaprakasha GK, Jagan Mohan Rao L, Sakariah KK. Improved HPLC method for the determination of curcumin, demethoxycurcumin, and bisdemethoxycurcumin. J Agric Food Chem 2002; 50(13): 3668-72.
[http://dx.doi.org/10.1021/jf025506a] [PMID: 12059141]

[62]   S JHV, A G, A Y, A Z, N A, P E. Antioxidant status, immune system, blood metabolites and carcass characteristic of broiler chickens fed turmeric rhizome powder under heat stress. Afr J Biotechnol 2012; 11(94): 16118-25.
[http://dx.doi.org/10.5897/AJB12.1986]

[63]   Kurkure NV, Pawar SP, Kognole SM, Bhandarkar AG, Ganorkar AG, Kalorey DR. Ameliorative effect of turmeric *Curcuma longa* in induced aflatoxicosis in cockerels. Indian J Vet Pathol 2000; 24: 26-8.

[64]   Sawale GK, Gosh RC, Ravikanth K, Maini S, Rekhe DS. Experimental mycotoxicosis in layer induced by ochratoxin A and its amelioration with herbomineral toxin binder 'Toxiroak'. Int J Poult Sci 2009; 8(8): 798-803.
[http://dx.doi.org/10.3923/ijps.2009.798.803]

[65]   Abbas RZ, Iqbal Z, Khan MN, Zafar MA, Zia MA. Anticoccidial activity of *Curcuma longa* L. in broilers. Braz Arch Biol Technol 2010; 53(1): 63-7.
[http://dx.doi.org/10.1590/S1516-89132010000100008]

[66]   Yarru LP, Settivari RS, Gowda NKS, Antoniou E, Ledoux DR, Rottinghaus GE. Effects of turmeric (*Curcuma longa*) on the expression of hepatic genes associated with biotransformation, antioxidant, and immune systems in broiler chicks fed aflatoxin. Poult Sci 2009; 88(12): 2620-7.
[http://dx.doi.org/10.3382/ps.2009-00204] [PMID: 19903961]

[67]   Tirupathi R. Effect of herbal preparations on the performance of broilers. M.V.Sc., Thesis submitted to Sri Venkateswara Veterinary University, Tirupathi. 2010.

[68]   Emadi M, Kermanshahi H. Effect of turmeric rhizome powder on immunity responses of broiler chickens. J Anim Vet Adv 2007; 6: 833-6.

[69]   Lee SH, Lillehoj HS, Jang SI, Lillehoj EP, Min W, Bravo DM. Dietary supplementation of young broiler chickens with Capsicum and turmeric oleoresins increases resistance to necrotic enteritis. Br J Nutr 2013; 110(5): 840-7.
[http://dx.doi.org/10.1017/S0007114512006083] [PMID: 23566550]

[70]   Garcea G, Berry DP, Jones DJL, *et al.* Consumption of the putative chemopreventive agent curcumin by cancer patients: assessment of curcumin levels in the colorectum and their pharmacodynamic consequences. Cancer Epidemiol Biomarkers Prev 2005; 14(1): 120-5.
[http://dx.doi.org/10.1158/1055-9965.120.14.1] [PMID: 15668484]

[71]   Jagetia GC, Aggarwal BB. "Spicing up" of the immune system by curcumin. J Clin Immunol 2007; 27(1): 19-35.
[http://dx.doi.org/10.1007/s10875-006-9066-7] [PMID: 17211725]

[72]   Antony S, Kuttan R, Kuttan G. Immunomodulatory activity of curcumin. Immunol Invest 1999; 28(5-6): 291-303.
[http://dx.doi.org/10.3109/08820139909062263] [PMID: 10574627]

[73]   Chandra RG, Chakraborty K, Nandy P, Moitra MN. Pros and cons of curcumin as bioactive phyto-compound for effective management of insect pests. Am Sci Res J Eng Technol Sci 2014; 7: 2313-4410.

[74]   Sagnou M, Mitsopoulou KP, Koliopoulos G, Pelecanou M, Couladouros EA, Michaelakis A. Evaluation of naturally occurring curcuminoids and related compounds against mosquito larvae. Acta Trop 2012; 123(3): 190-5.
[http://dx.doi.org/10.1016/j.actatropica.2012.05.006] [PMID: 22634203]

CHAPTER 8

Oregano Essential Oils

Youssef A. Attia[1,2,*], **Mohamed E. Abd El-Hack**[3], **Ayman E. Taha**[4], **Mohamed A. AlBanoby**[5], **Adel D. Al-qurashi**[1], **Asmaa F. Khafaga**[6], **Vincenzo Tufarelli**[7] and **Maria Cristina De Oliveira**[8]

[1] *Sustainable Agriculture Production Research Group, Agriculture Department, Faculty of Environmental Sciences, King Abdulaziz University, Jeddah-21589, Saudi Arabia*

[2] *Animal and Poultry Production Department, Faculty of Agriculture, Damanhour University, Damanhour-22713, Egypt*

[3] *Poultry Department, Faculty of Agriculture, Zagazig University, Zagazig-44519, Egypt*

[4] *Department of Animal Husbandry and Animal Wealth Development, Faculty of Veterinary Medicine, Alexandria University, Apis, Alexandria, 21944, Egypt*

[5] *Al-Shamel Animal Feed Factory, Industrial Area, Hail-55411, Saudi Arabia*

[6] *Department of Pathology, Faculty of Veterinary Medicine, Alexandria University, Apis, Alexandria, 21944, Egypt*

[7] *Department of Precision and Regenerative Medicine and Jonian Area (DiMePRe-J), Section of Veterinary Science and Animal Production, University of Bari 'Aldo Moro' s.p. Casamassima km 3, 70010 Valenzano, Italy*

[8] *University of Rio Verde, Faculty of Veterinary Medicine, Rio Verde, GO, Brazil*

Abstract: Essential oils (EOs) are aromatic products made from a combination of components extracted from plant materials used in food, cosmetics, and medicine, among several other applications. EOs are extracted using various extraction methods from the bark, seeds, leaves, peel, buds, flowers, and other components of medicinal plants. Techniques used to extract EO include steam distillation, solvent-assisted extraction, hydrodistillation, ultrasonic extraction, supercritical fluid extraction, and solvent-free microwave extraction. EO affects the intestinal health and growth efficiency of different animal species. EO has been reported to improve pancreatic amylase, trypsin, and maltase levels and increase digestibility. EO has antioxidant action, lowers lipid oxidation in meat, and enhances shelf-life. The present chapter summarizes some of the beneficial effects of oregano EO on poultry production and health.

*　**Corresponding author Youssef A. Attia:** Sustainable Agriculture Production Research Group, Agriculture Department, Faculty of Environmental Sciences, King Abdulaziz University, Jeddah-22713, Saudi Arabia & Animal and Poultry Production Department, Faculty of Agriculture, Damanhour University, Damanhour-22516, Egypt; E-mail: yaattia@kau.edu.sa

Keywords: Essential oil, Gut health, Growth promoter, Livestock, Oxidative stress.

INTRODUCTION

For several decades, antibiotics have been used as growth and health promoters in animal production [1]. Feed additives can also improve animal performance. However, some additives may be added as substitutes for antibiotics to prevent disease. The effectiveness of feed additives varies significantly [2]. One such additive is essential oils (EO), which have broad effects on different parameters of poultry under various management and environmental conditions [3, 4].

Essential oils are aromatic products made from a combination of components extracted from plants. These plants' products are utilized in food, cosmetics, and medicine, among several other applications [5], and the EO are extracted using a variety of extraction methods from the bark, seeds, leaves, peels, buds, flowers, and other components of medicinal plants [6]. The methods used to extract EO include steam distillation, solvent-assisted extraction, hydro distillation, ultrasonic extraction, supercritical fluid extraction, and solvent-free microwave extraction [7].

The primary ingredients of each oil vary [3]. Sesquiterpenes, monoterpenes, aldehydes, alcohols, esters, and ketones constitute some of the various volatile compounds that constitute essential oils. They have been linked to the development of plant defenses against pathogens, herbivores, fungi, and bacteria [8]. Essential oils are used as feed additives in animal production because of their antiviral, antimicrobial, anti-parasitic, and antifungal properties. Additionally, EO plays a role in appetite stimulation, enhancing the secretion of food digestion enzymes and activating the immune response [9].

EO affects the intestinal health and growth efficiency of pigs [10, 11], fishes [12 - 14], rabbits [15, 16], poultries [17 - 19], and ruminants [20 - 22]. Jang *et al.* [23] showed that broilers fed a combination of EO exhibited higher pancreatic trypsin, amylase, and maltase (digestive enzyme activity) in the gut compared to birds not fed EO.

Therefore, greater nutritional absorption may result from a combination of the antibacterial action of EO and an increase in digestive secretions. Changes in the microscopic anatomy of the small intestine and shifts in the gut microbiota are the two primary mechanisms by which EO affect gut health. When these two effects are combined, animals can better fight diseases from pathogenic organisms [24, 25].

Little information is available, particularly on the effect of EO on intestinal histology in poultry. Because of their antioxidant action, EO, including oregano, sage, carvacrol, thymol, and rosemary, are also known to reduce lipid oxidation in meat [25 - 27]. The present chapter summarizes some of the beneficial effects of oregano EO on poultry production and health.

OREGANO ESSENTIAL OILS

Origanum vulgare, also known as wild marjoram (family *Lamiaceae*), is native to North America, originally from the Mediterranean, and has a long history of use as a food and medicinal plant. Its name is derived from the Greek, origanon, a perennial herb that can be as tall as 80 cm, with dark oval, fragrant leaves, and spikes of pink, white, or purple flowers [28]. Oregano essential oil (OEO) contains high concentrations of carvacrol, p-cymene, c-terpinene, limonene, terpinene, ocimene, caryophyllene, β-bisabolene, linalool, and 4-terpineol [29].

Chemical and Bioactive Constituents

Alkhafaji and Jayashankar [30] discovered 66 distinct compounds in oregano oil, which accounted for 100% of the makeup. Oregano oil has been shown to have a 187.94 mg KOH/g saponification value, a 6.22 mg/g peroxide value, and a 3.645 mg/g P-anisidine value.

Carvacrol and thymol represent most of the phenolic chemicals found in oregano oil, accounting for more than 50% of it [31]. Terpenes, typically mono- and sesquiterpenes, are the main elements of EO and carvacrol, while thymol, -terpinene, and p-cymene are the primary terpenes found in oregano [32, 33], as shown in Fig. (**1**).

The chemotype of a species is determined by the ratio of these and other essential oil (EO) components. Typically, the main EO components are used to designate the chemotype, such as carvacrol, thymol, -citronellol, and 1, 8-cineole. For instance, Lukas *et al.* [34] identified three chemotypes of *O. vulgare* based on the amounts of cymyl-, sabinyl-, and linalool/linalyl acetate in extracts of 502 distinct plants from 17 different nations.

Mechanism of Biological Activity

Herbal plants can affect animal health and production in several ways. Many plants have antioxidant, antimicrobial, anti-inflammatory, and antiproliferative properties [32, 35, 36] that enhance digestibility, immunity, and intestinal health [19, 37].

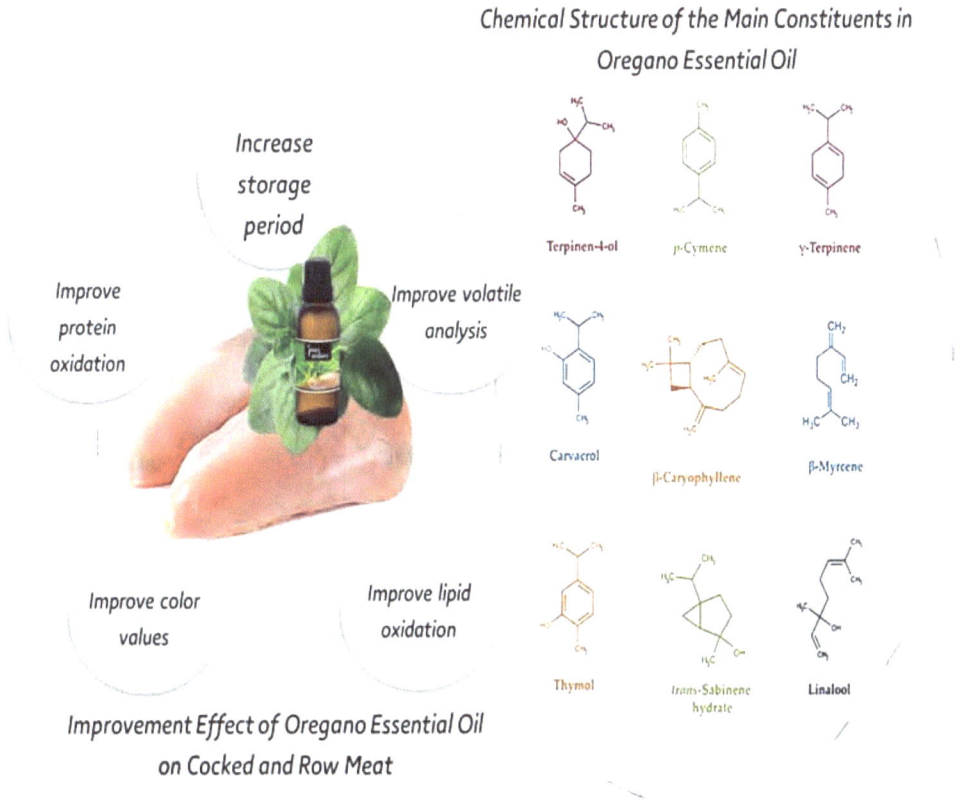

Fig. (1). Chemical structure of the main constituents of oregano essential oils and their effects on meat quality.

The beneficial effects of OEO are summarized in Fig. (**2**). Supplementing poultry with EO, such as OEO, has been proven to offer numerous advantages, including improved growth, feed conversion ratio, gastrointestinal morphology, blood lipid profile, antioxidant capacity, and immune status, either humeral or cellular [17, 18, 38, 39]. Dietary oregano supplementation promotes the performance of Hy-Line Brown pullets [40].

Oregano Essential Oil Effects

Poultry Growth

The goal of success in any commercial poultry flock is to improve the growth performance of birds, according to the breeding guide for each species. According to Ertas *et al.* [41] OEO at 200 mg/kg for 3 weeks increased broiler growth performance, improved feed conversion rate (FCR), and increased appetite and body weight gain (BWG) compared to broilers fed diets without OEO.

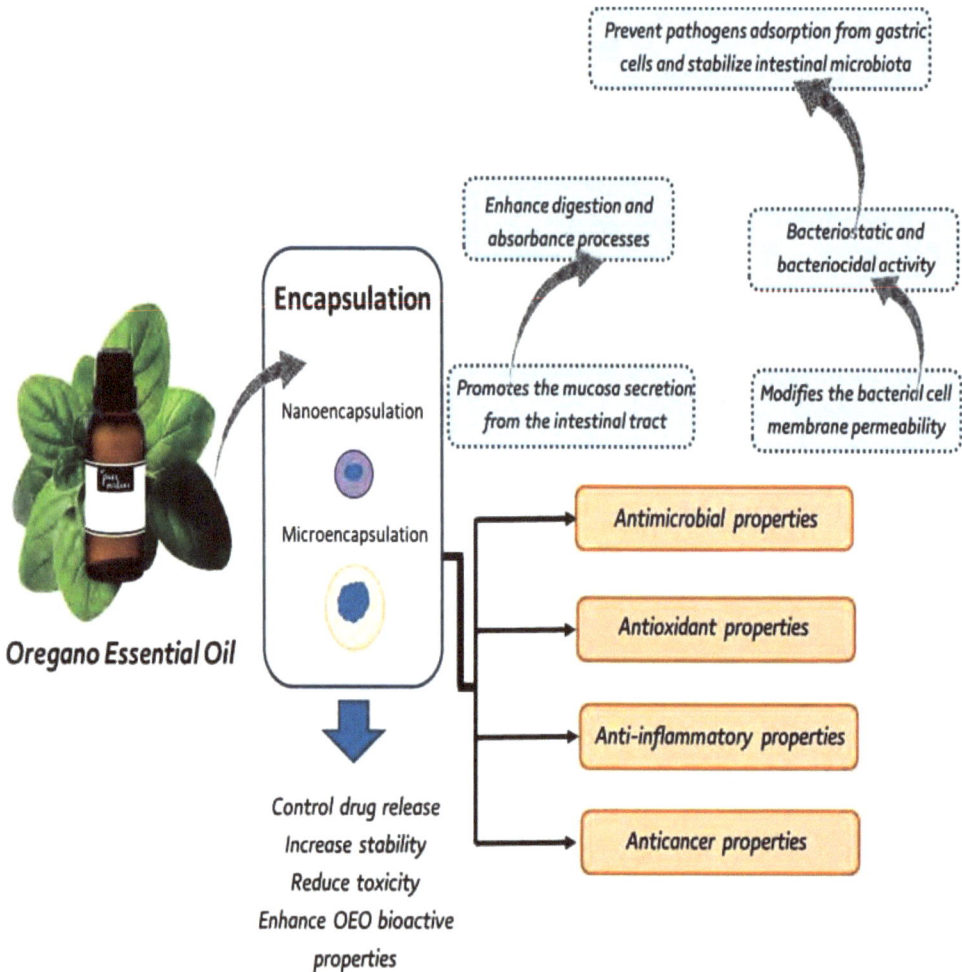

Fig. (2). Beneficial aspects of oregano essential oil.

Peng *et al.* [42] studied the effect of OEO (300 and 600 mg/kg) on broiler efficiency for 42 d and compared it to using 8 mg/kg of avilamycin. Both doses of OEO increased daily feed consumption and BWG compared to the control group. Moreover, Namdeo *et al.* [43] showed that broilers fed myrtle leaf EO, oregano, sage leaf, laurel leaf, citrus peel, and fennel seeds had better FCR.

One explanation may be that OEO promotes digestibility by stimulating enzyme secretion in the gut, as hypothesized by Ghazi *et al.* [44] who supplemented broilers with OEO under heat stress. Another explanation could be associated with the increased nutrient metabolism, as suggested by Reyer *et al.* [45], who supplemented day-old broilers with a dietary EO blend of star anise, rosemary, thyme, and oregano.

Johnson *et al.* [40] demonstrated that birds supplemented with OEO had higher daily BWG but no impact on feed intake (FI) compared to control birds. This may be because OEO increases digestive health through better nutrient utilization, as described by Zhang *et al.* [21]. Average body weight, livability, and daily FI were improved by supplementation with OEO in the diets of yellow-feathered chickens, with no effect on FCR [46].

Studying the use of OEO in Japanese quails, Farouk *et al.* [47] reported that weight gain increased despite the lower FI in birds fed a diet containing 150 mg/kg OEO to the control group and in birds receiving 300 mg/kg of dietary OEO.

Immunity and Antimicrobial Effects

The main objectives of poultry production are to boost immunity to reduce the risk of infection by numerous diseases, immunization failures, illnesses that weaken the immune system, and the inappropriate use of antibiotics.

Resistance to many infections is greatly influenced by the host phagocyte system, cellular immunity, and humoral defenses [48, 49]. The phenolic components of OEO may increase the humoral and cellular immunological responses of broiler chickens [50]. Essential oils comprise many compounds that give the oil its distinct characteristics; these active substances share the antibacterial, antiviral, and antifungal characteristics of their parent plant [51].

Birds fed OEO at 300 mg/kg had higher levels of ileal IgA and plasma IgG, and lower mucosal ileal TNF content than the control group [46]. Rostro-Alanis *et al.* [52] found that OEO showed an antimicrobial effect when assessing the zone of inhibition against *Staphylococcus aureus, Listeria monocytogenes, Salmonella typhi,* and *Candida albicans.* Adding clove or oregano EO to feed may enhance broiler performance and mitigate some of the effects of coccidial damage on meat quality and gut integrity [21].

The phenolic content of EO, such as resveratrol, eugenol cinnamaldehyde, curcumin, thymol, and carvacrol, which are derived from the herbs thyme, oregano, turmeric, clove, and cinnamon, may be the cause of their antibacterial action [53, 54]. Antiviral effects occur in two ways: first, they reduce viral production and second, they reduce viral DNA or RNA synthesis [55].

Abouelezz *et al.* [56] fed ducklings with OEO at 150 and 300 mg/kg diet and found fewer coliform bacteria in the cecum. Extended-spectrum beta-lactamase (ESBL)-generating *E. coli* bacteria were treated with OEO and antibiotics. By reducing the effective dose of antibiotics, OEO reduces the adverse effects of antibiotics on ESBL-producing *E. coli* [57]. Carvacrol, thymol, trans-cinnamaldehyde, and tetrasodium pyrophosphate significantly decreased the populations of *E. coli* and *S. typhi* in chicken breast [58].

The formation of *C. perfringens* spores in ground turkey during chilling was also inhibited by carvacrol, cinnamaldehyde, oregano oil, and thymol [59].

It is well known that the hydrophobicity of herbal constituents, which affects the permeability of cell membranes and cell homeostasis and results in the loss of cellular components and even cell death, is a critical component of their antibacterial activity [60]. This property enables herbal compounds to penetrate the lipid layers in the cell walls of bacteria and mitochondria and disturb the cell structure [61]. Lambert *et al.* [62] showed that two significant components of oregano essential oil, carvacrol and thymol, had a synergistic effect when tested against *S. aureus* and *P. aeruginosa*.

Antioxidant and Lipid Profile

Lipid peroxidation refers to the defense against the deterioration of cell membranes. Additionally, harmful processes of lipid peroxidation can generate free radicals. Malondialdehyde (MDA) is the primary biomarker for the oxidative process.

Indicators of antioxidant activity include superoxide dismutase (SOD), glutathione peroxidase (GSH-Px), and total antioxidant capacity (TAC) [63]. Potent antioxidant substances in OEO, such as carvacrol and thymol, may improve an animal's resistance to oxidative stress [64]. Botsoglou *et al.* [65] observed a significant antioxidant effect in the broiler tissues after adding 0.05 to 0.1g/kg of oregano extract to broiler chicken feed. According to Luna *et al.*, thymol and carvacrol reduce lipid oxidation in the diet more than any other synthetic substance [66]. In a study by Johnson *et al.* [40], MDA levels were lower in the OEO group than in the control group, but glutathione and TAC levels were higher in the OEO group.

In broiler hens fed 150 mg/kg OEO, Ruan *et al.* [46] found increased GSH-Px and TAC activity in the plasma, jejunum, and ileum. Mueller *et al.* [67] administered

150 mg/kg of OEO, which contains 65% carvacrol, to broiler chicks from 15 - 35 days of age, and noted improvement in the activities of GSH-Px and SOD in the jejunum but decreased hepatic activity without influencing lipid peroxidation.

Thymol and oregano enhanced the oxidative stability of eggs [68, 69]. Giannenas *et al.* [70] and Oral *et al.* examined the extension of the shelf life of packaged fresh chicken [71]. The authors cited that OEO extended the product's shelf life by two days.

The liver primarily controls lipid metabolism [72]. Lipid metabolism included triglyceride, cholesterol, and high-density lipoprotein (HDL) levels. High levels of triglycerides may indicate hepatocyte injury, as they are frequently absorbed from the diet in the intestines and can be retained in the liver [73].

Cholesterol can be produced by the liver or removed from the blood if in excess [74]. According to Ouimet *et al.* [75], high-density lipoproteins (HDL) can be produced in the liver and move extra cholesterol from the bloodstream and tissues to the liver.

Johnson *et al.* [40] discovered that using OEO in diets for layers at 11 and 17 weeks of age reduced triglyceride levels and increased HDL levels. At 11 weeks of age, cholesterol levels in the OEO group were higher than those in the control group. At 17 weeks of age, the cholesterol levels were not considerably different.

Abo Ghanima *et al.* [76] reported that thymol, one of the active compounds in OEO, reduces cholesterol synthesis, suggesting that the functional elements in OEO may influence the mechanisms of lipid metabolism.

According to Migliorini *et al.* [77], an increase in serum triglyceride levels was observed on day 84 in hens receiving 200 mg of OEO per kg of feed, which was consistent with the findings of Bolukbasi *et al.* [78] in broilers supplemented with thyme EO. These authors claim that this fact results from an increase in protein, fat, and carbohydrate metabolism linked to an increase in the concentration of triglycerides [79].

Intestinal Health

Johnson *et al.* [40] concluded that adding OEO to broilers resulted in increased weights and percentages of the entire gastrointestinal tract (GIT), gizzard, and proventriculus but had no effect on the weights of the liver or spleen. According to this study, birds that consumed OEO had larger GIT weights than those in the control group. A possible explanation for the increase in proventriculus weight may be attributed to EO supplementation, which increased specific enzymatic activity, as described by Hashemipour *et al.* [80].

The lack of differences at younger ages may be because food is not stored in this organ but passes through it quickly [81, 82]. Dietary OEO supplementation affected duodenal enzyme activity in broilers on day 42 with increased trypsin activity. According to Zhang *et al.* [21], supplementation with antibiotics and an OEO combination improved chymotrypsin, lipase, and amylase activity in broilers on day 42. Moreover, these findings suggested that OEO administration enhanced the production of digestive enzymes, increasing the digestibility of nutrients, and ultimately improving broiler growth performance.

In the duodenum on day 42, broilers fed OEO had a higher villus height (VH) and lower crypt depth (CD), and on days 21 and 42, higher VH/CD in the duodenum, jejunum, and ileum [21]. Similar reports were cited by Ding *et al.* [83], who reported a significant reduction in CD and increase in VH/CD in the jejunum of ducks. Additionally, supplementing the basal diet with 600 mg/kg OEO increased VH in laying hens [84].

According to Behnamifar *et al.* [85], oregano extract enhances VH in the ileum of quail, resulting in a valuable and low-risk poultry additive. The efficiency of digestion and absorption may increase with higher VH and VH/CD ratios in the jejunum [86]. Windisch *et al.* [87] hypothesized that the antioxidant properties of herbal essential oils are responsible for improving intestinal morphology. There are approximately 300 commercially available EOs; the inconsistent results in research could be caused by the fact that the chemical components in oils can be modified by species, climatic circumstances, harvest time, and plant parts [88].

CONCLUDING AND REMARKS

Oregano essential oil has demonstrated significant potential as a natural feed additive for poultry production. The active compounds in OEO, particularly carvacrol and thymol, have antimicrobial, antioxidant, and growth-promoting properties. Studies have shown that OEO supplementation can improve productive performance, feed conversion, and intestinal health. It enhances immune function, reduces pathogenic bacteria, and increases the antioxidant capacity in poultry. OEO also positively affected lipid metabolism and meat quality. While results can vary based on dosage and specific oil composition, the overall evidence suggests that OEO is a promising alternative to antibiotics in poultry feed. Further research is needed to optimize its application and understand its mechanisms of action completely; however, OEO shows great promise as a natural, multi-functional feed additive for improving poultry health and production.

IMPLICATIONS

- Numerous studies have examined the nature and applications of essential oils in poultry feeds. The active ingredients in OEOs have antibacterial, antioxidant, and toxic effects at high doses in laboratory studies. The antioxidant capabilities of OEO enhance animal performance, gut health, and other characteristics of growth and productivity.
- The benefits of oregano essential oil (OEO) in poultry production imply that it could serve as a viable natural alternative to antibiotic growth promoters.
- The multifaceted effects of OEO on growth performance, intestinal health, immunity, and meat quality suggest that its incorporation into poultry diets may lead to more sustainable and consumer-friendly production practices.
- The variability in OEO composition and efficacy across studies indicates the need for standardization and further research to optimize its use in commercial poultry operations and organic poultry production.
- The potential of OEO to enhance poultry health and productivity while addressing concerns about antibiotic resistance implies that it could play a significant role in future poultry nutrition and management strategies.

REFERENCES

[1]   Oliveira MC, Attia YA, Khafaga AF, *et al. Withania somnifera* is a phytoherbal growth promoter for broiler farming. Ann Anim Sci 2023. Ahead of print [https://doi.org/10.2478/aoas-2023-0045].

[2]   Abo Ghanima MM, Abd El-Hack ME, Al-Otaibi AM, *et al.* Growth performance, liver and kidney functions, blood hormonal profile, and economic efficiency of broilers fed different levels of threonine supplementation during feed restriction. Poult Sci 2023; 102(8): 102796.
[http://dx.doi.org/10.1016/j.psj.2023.102796] [PMID: 37321031]

[3]   Zhai H, Liu H, Wang S, Wu J, Kluenter AM. Potential of essential oils for poultry and pigs. Anim Nutr 2018; 4(2): 179-86.
[http://dx.doi.org/10.1016/j.aninu.2018.01.005] [PMID: 30140757]

[4]   Sadhasivam S, Shapiro OH, Ziv C, Barda O, Zakin V, Sionov E. Synergistic inhibition of mycotoxigenic fungi and mycotoxin production by combination of pomegranate peel extract and azole fungicide. Front Microbiol 2019; 10: 1919.
[http://dx.doi.org/10.3389/fmicb.2019.01919] [PMID: 31481948]

[5]   Kant R, Kumar A. Review on essential oil extraction from aromatic and medicinal plants: Techniques, performance and economic analysis. Sustain Chem Pharm 2022; 30: 100829.
[http://dx.doi.org/10.1016/j.scp.2022.100829]

[6]   Tohidi B, Rahimmalek M, Trindade H. Review on essential oil, extracts composition, molecular and phytochemical properties of *Thymus* species in Iran. Ind Crops Prod 2019; 134: 89-99.
[http://dx.doi.org/10.1016/j.indcrop.2019.02.038]

[7]   Belhachat D, Mekimene L, Belhachat M, Ferradji A, Aid F. Application of response surface methodology to optimize the extraction of essential oil from ripe berries of *Pistacia lentiscus* using ultrasonic pretreatment. J Appl Res Med Aromat Plants 2018; 9: 132-40.
[http://dx.doi.org/10.1016/j.jarmap.2018.04.003]

[8]   Harkat-Madouri L, Asma B, Madani K, *et al.* Chemical composition, antibacterial and antioxidant activities of essential oil of *Eucalyptus globulus* from Algeria. Ind Crops Prod 2015; 78: 148-53.
[http://dx.doi.org/10.1016/j.indcrop.2015.10.015]

[9] Krishan G, Narang A. Use of essential oils in poultry nutrition: A new approach. J Adv Vet Anim Res 2014; 1(4): 156-62.
[http://dx.doi.org/10.5455/javar.2014.a36]

[10] Ruzauskas M, Bartkiene E, Stankevicius A, *et al.* The influence of essential oils on gut microbial profiles in pigs. Animals (Basel) 2020; 10(10): 1734.
[http://dx.doi.org/10.3390/ani10101734] [PMID: 32987688]

[11] Zhao BC, Wang TH, Chen J, *et al.* Effects of dietary supplementation with a carvacrol–cinnamaldehyde–thymol blend on growth performance and intestinal health of nursery pigs. Porcine Health Manag 2023; 9(1): 24.
[http://dx.doi.org/10.1186/s40813-023-00317-x] [PMID: 37221604]

[12] Ceppa F, Faccenda F, De Filippo C, *et al.* Influence of essential oils in diet and life-stage on gut microbiota and fillet quality of rainbow trout (*Oncorhynchus mykiss*). Int J Food Sci Nutr 2018; 69(3): 318-33.
[http://dx.doi.org/10.1080/09637486.2017.1370699] [PMID: 28859525]

[13] Souza EMD, Souza RCD, Melo JFB, *et al. Cymbopogon flexuosus* essential oil as an additive improves growth, biochemical and physiological responses and survival against *Aeromonas hydrophila* infection in Nile tilapia. An Acad Bras Cienc 2020; 92 (Suppl. 1): e20190140.
[http://dx.doi.org/10.1590/0001-3765202020190140] [PMID: 32638863]

[14] Zaminhan-Hassemer M, Zagolin GB, Perazza CA, *et al.* Adding an essential oil blend to the diet of juvenile Nile tilapia improves growth and alters the gut microbiota. Aquaculture 2022; 560: 738581.
[http://dx.doi.org/10.1016/j.aquaculture.2022.738581]

[15] Li C, Niu J, Liu Y, Li F, Liu L. The effects of oregano essential oil on production performance and intestinal barrier function in growing Hyla rabbits. Ital J Anim Sci 2021; 20(1): 2165-73.
[http://dx.doi.org/10.1080/1828051X.2021.2005471]

[16] Placha I, Bacova K, Zitterl-Eglseer K, *et al.* Thymol in fattening rabbit diet, its bioavailability and effects on intestinal morphology, microbiota from caecal content and immunity. J Anim Physiol Anim Nutr (Berl) 2022; 106(2): 368-77.
[http://dx.doi.org/10.1111/jpn.13595] [PMID: 34156121]

[17] Adewole DI, Oladokun S, Santin E. Effect of organic acids–essential oils blend and oat fiber combination on broiler chicken growth performance, blood parameters, and intestinal health. Anim Nutr 2021; 7(4): 1039-51.
[http://dx.doi.org/10.1016/j.aninu.2021.02.001] [PMID: 34738034]

[18] Su G, Wang L, Zhou X, *et al.* Effects of essential oil on growth performance, digestibility, immunity, and intestinal health in broilers. Poult Sci 2021; 100(8): 101242.
[http://dx.doi.org/10.1016/j.psj.2021.101242] [PMID: 34174571]

[19] Hosseinzadeh S, Shariatmadari F, Karimi Torshizi MA, Ahmadi H, Scholey D. *Plectranthus amboinicus* and rosemary (*Rosmarinus officinalis* L.) essential oils effects on performance, antioxidant activity, intestinal health, immune response, and plasma biochemistry in broiler chickens. Food Sci Nutr 2023; 11(7): 3939-48.
[http://dx.doi.org/10.1002/fsn3.3380] [PMID: 37457190]

[20] Sun J, Cheng Z, Zhao Y, Wang Y, Wang H, Ren Z. Influence of increasing levels of oregano essential oil on intestinal morphology, intestinal flora and performance of Sewa sheep. Ital J Anim Sci 2022; 21(1): 463-72.
[http://dx.doi.org/10.1080/1828051X.2022.2048208]

[21] Zhang LY, Peng QY, Liu YR, *et al.* Effects of oregano essential oil as an antibiotic growth promoter alternative on growth performance, antioxidant status, and intestinal health of broilers. Poult Sci 2021; 100(7): 101163.
[http://dx.doi.org/10.1016/j.psj.2021.101163] [PMID: 34082177]

[22]   Jia L, Wu J, Lei Y, *et al.* Oregano essential oils mediated intestinal microbiota and metabolites and improved growth performance and intestinal barrier function in sheep. Front Immunol 2022; 13: 908015.
[http://dx.doi.org/10.3389/fimmu.2022.908015] [PMID: 35903106]

[23]   Jang IS, Ko YH, Kang SY, Lee CY. Effect of a commercial essential oil on growth performance, digestive enzyme activity and intestinal microflora population in broiler chickens. Anim Feed Sci Technol 2007; 134(3-4): 304-15.
[http://dx.doi.org/10.1016/j.anifeedsci.2006.06.009]

[24]   Basmacıoğlu-Malayoğlu H, Özdemir P, Bağriyanik HA. Influence of an organic acid blend and essential oil blend, individually or in combination, on growth performance, carcass parameters, apparent digestibility, intestinal microflora and intestinal morphology of broilers. Br Poult Sci 2016; 57(2): 227-34.
[http://dx.doi.org/10.1080/00071668.2016.1141171] [PMID: 26785140]

[25]   Negera M, Washe AP. Use of natural dietary spices for reclamation of food quality impairment by aflatoxin. J Food Qual 2019; 2019: 1-10.
[http://dx.doi.org/10.1155/2019/4371206]

[26]   Brenes A, Roura E. Essential oils in poultry nutrition: Main effects and modes of action. Anim Feed Sci Technol 2010; 158(1-2): 1-14.
[http://dx.doi.org/10.1016/j.anifeedsci.2010.03.007]

[27]   Chaves Lobón N, Ferrer de la Cruz I, Alías Gallego JC. Autotoxicity of diterpenes present in leaves of *Cistus ladanifer* L. Plants 2019; 8(2): 27.
[http://dx.doi.org/10.3390/plants8020027] [PMID: 30678267]

[28]   Kumar V, Marković T, Emerald M, Dey A. Herbs: composition and dietary importance. In: Caballero B, Finglas PM, Toldrá F, Eds. Encyclopedia of food and health. London: Academic Press 2016; pp. 332-7.
[http://dx.doi.org/10.1016/B978-0-12-384947-2.00376-7]

[29]   Leyva-López N, Gutiérrez-Grijalva E, Vazquez-Olivo G, Heredia J. Essential oils of oregano: biological activity beyond their antimicrobial properties. Molecules 2017; 22(6): 989.
[http://dx.doi.org/10.3390/molecules22060989] [PMID: 28613267]

[30]   Hadi Alkhafaji RT, Jayashankar M. Physicochemical properties and inhibitory effects of oregano oil against uropathogenic. Pharmacognosy Res 2022; 14(3): 328-32.
[http://dx.doi.org/10.5530/pres.14.3.48]

[31]   Han F, Ma GG, Yang M, *et al.* Chemical composition and antioxidant activities of essential oils from different parts of the oregano. J. Zhejiang Univ. Sci 2017; B. 18(1): 79.
[http://dx.doi.org/10.1631/jzus.B1600377]

[32]   Masyita A, Sari RM, Astuti AD, *et al.* Terpenes and terpenoids as main bioactive compounds of essential oils, their roles in human health and potential application as natural food preservatives. Food Chem 2022; X 13: 100217.
[http://dx.doi.org/10.1016/j.fochx.2022.100217]

[33]   Cao G, Liu J, Liu H, *et al.* Integration of network pharmacology and molecular docking to analyse the mechanism of action of oregano essential oil in the treatment of bovine mastitis. Vet Sci 2023; 10(5): 350.
[http://dx.doi.org/10.3390/vetsci10050350] [PMID: 37235433]

[34]   Lukas B, Schmiderer C, Novak J. Essential oil diversity of European *Origanum vulgare* L. (Lamiaceae). Phytochemistry 2015; 119: 32-40.
[http://dx.doi.org/10.1016/j.phytochem.2015.09.008] [PMID: 26454793]

[35]   Tlili H, Hanen N, Ben Arfa A, *et al.* Biochemical profile and *in vitro* biological activities of extracts from seven folk medicinal plants growing wild in southern Tunisia. PLoS One 2019; 14(9): e0213049.

[http://dx.doi.org/10.1371/journal.pone.0213049] [PMID: 31527869]

[36] Romeiras MM, Essoh AP, Catarino S, *et al.* Diversity and biological activities of medicinal plants of Santiago island (Cabo Verde). Heliyon 2023; 9(4): e14651.
[http://dx.doi.org/10.1016/j.heliyon.2023.e14651] [PMID: 37009246]

[37] Elbaz AM, Ashmawy ES, Salama AA, Abdel-Moneim AME, Badri FB, Thabet HA. Effects of garlic and lemon essential oils on performance, digestibility, plasma metabolite, and intestinal health in broilers under environmental heat stress. BMC Vet Res 2022; 18(1): 430.
[http://dx.doi.org/10.1186/s12917-022-03530-y] [PMID: 36503512]

[38] Iqbal H, Rahman A, Khanum S, *et al.* Effect of essential oil and organic acid on performance, gut health, bacterial count and serological parameters in broiler. Braz. J. Poult. Sci 2021; 23(3): eRBCA-2021-1443.
[http://dx.doi.org/10.1590/1806-9061-2021-1443]

[39] Gholami-Ahangaran M, Ahmadi-Dastgerdi A, Azizi S, Basiratpour A, Zokaei M, Derakhshan M. Thymol and carvacrol supplementation in poultry health and performance 2022; Vet. Med. Sci. 8(1): 267-288.
[http://dx.doi.org/10.1002/vms3.663]

[40] Johnson AM, Anderson G, Arguelles-Ramos M, Ali AAB. Effect of dietary essential oil of oregano on performance parameters, gastrointestinal traits, blood lipid profile, and antioxidant capacity of laying hens during the pullet phase. Front Anim Sci 2022; 3: 1072712.
[http://dx.doi.org/10.3389/fanim.2022.1072712]

[41] Talat Guler , Mehmet Ciftci , Dalkilic B, Dalkilic B, Simsek UG. The effect of an essential oil mix derived from oregano, clove and anise on broiler performance. Int J Poult Sci 2005; 4(11): 879-84.
[http://dx.doi.org/10.3923/ijps.2005.879.884]

[42] Peng QY, Li JD, Li Z, Duan ZY, Wu YP. Effects of dietary supplementation with oregano essential oil on growth performance, carcass traits and jejunal morphology in broiler chickens. Anim Feed Sci Technol 2016; 214: 148-53.
[http://dx.doi.org/10.1016/j.anifeedsci.2016.02.010]

[43] Namdeo S, Baghel RPS, Nayak S, *et al.* Essential oils: a potential substitute to antibiotics growth promoter in broiler diet. J Entomol Zool Stud 2020; 8(4): 1643-9.

[44] Ghazi S, Amjadian T, Norouzi S. Single and combined effects of vitamin C and oregano essential oil in diet, on growth performance, and blood parameters of broiler chicks reared under heat stress condition. Int J Biometeorol 2015; 59(8): 1019-24.
[http://dx.doi.org/10.1007/s00484-014-0915-4] [PMID: 25336108]

[45] Reyer H, Zentek J, Männer K, *et al.* Possible molecular mechanisms by which an essential oil blend from star anise, rosemary, thyme, and oregano and saponins increase the performance and ileal protein digestibility of growing broilers. J Agric Food Chem 2017; 65(32): 6821-30.
[http://dx.doi.org/10.1021/acs.jafc.7b01925] [PMID: 28722406]

[46] Ruan D, Fan Q, Fouad AM, *et al.* Effects of dietary oregano essential oil supplementation on growth performance, intestinal antioxidative capacity, immunity, and intestinal microbiota in yellow-feathered chickens. J Anim Sci 2021; 99(2): skab033.
[http://dx.doi.org/10.1093/jas/skab033] [PMID: 33544855]

[47] Farouk SM, Yusuf MS, El Nabtiti AAS, Abdelrazek HMA. Effect of oregano essential oil supplementation on performance, biochemical, hematological parameters and intestinal histomorphometry of Japanese quail (*Coturnix coturnix Japonica*). Vet Res Forum 2020; 11(3): 219-27.
[http://dx.doi.org/10.30466/vrf.2019.97574.2325] [PMID: 33133458]

[48] Linnerz T, Hall CJ. The diverse roles of phagocytes during bacterial and fungal infections and sterile inflammation: lessons from zebrafish. Front Immunol 2020; 11: 1094.
[http://dx.doi.org/10.3389/fimmu.2020.01094] [PMID: 32582182]

[49]   Ahmad HI, Jabbar A, Mushtaq N, *et al.* Immune tolerance *vs.* immune resistance: the interaction between host and pathogens in infectious diseases. Front Vet Sci 2022; 9: 827407.
[http://dx.doi.org/10.3389/fvets.2022.827407] [PMID: 35425833]

[50]   Acamovic T, Brooker JD. Biochemistry of plant secondary metabolites and their effects in animals. Proc Nutr Soc 2005; 64(3): 403-12.
[http://dx.doi.org/10.1079/PNS2005449] [PMID: 16048675]

[51]   Bakkali F, Averbeck S, Averbeck D, Idaomar M. Biological effects of essential oils – A review. Food Chem Toxicol 2008; 46(2): 446-75.
[http://dx.doi.org/10.1016/j.fct.2007.09.106] [PMID: 17996351]

[52]   Rostro-Alanis MJ, Báez-González J, Torres-Alvarez C, Parra-Saldívar R, Rodriguez-Rodriguez J, Castillo S. Chemical composition and biological activities of oregano essential oil and its fractions obtained by vacuum distillation. Molecules 2019; 24(10): 1904.
[http://dx.doi.org/10.3390/molecules24101904] [PMID: 31108897]

[53]   Veldhuizen EJA, Tjeerdsma-van Bokhoven JLM, Zweijtzer C, Burt SA, Haagsman HP. Structural requirements for the antimicrobial activity of carvacrol. J Agric Food Chem 2006; 54(5): 1874-9.
[http://dx.doi.org/10.1021/jf052564y] [PMID: 16506847]

[54]   Alagawany MM, Farag MR, Dhama K. Nutritional and biological effects of turmeric (*Curcuma longa*) supplementation on performance, serum biochemical parameters and oxidative status of broiler chicks exposed to endosulfan in the diets. Asian J Anim Vet Adv 2015; 10(2): 86-96.
[http://dx.doi.org/10.3923/ajava.2015.86.96]

[55]   Dwivedy AK, Singh VK, Prakash B, Dubey NK. Nanoencapsulated *Illicium verum* Hook.f. essential oil as an effective novel plant-based preservative against aflatoxin B1 production and free radical generation. Food Chem Toxicol 2018; 111: 102-13.
[http://dx.doi.org/10.1016/j.fct.2017.11.007] [PMID: 29126800]

[56]   Abouelezz K, Abou-Hadied M, Yuan J, *et al.* Nutritional impacts of dietary oregano and Enviva essential oils on the performance, gut microbiota and blood biochemicals of growing ducks. Animal 2019; 13(10): 2216-22.
[http://dx.doi.org/10.1017/S1751731119000508] [PMID: 30914073]

[57]   Gholami-Ahangaran M, Ahmadi-Dastgerdi A, Karimi-Dehkordi M. Thymol and carvacrol; as antibiotic alternative in green healthy poultry production. Plant Biotechnol 2020; Persa 2(1): 22-25.

[58]   Lacroix M, Chiasson F. The influence of MAP condition and active compounds on the radiosensitization of *Escherichia coli* and *Salmonella typhi* present in chicken breast. Radiat Phys Chem 2004; 71(1-2): 69-72.
[http://dx.doi.org/10.1016/j.radphyschem.2004.04.059]

[59]   Juneja VK, Friedman M. Carvacrol, cinnamaldehyde, oregano oil, and thymol inhibit *Clostridium perfringens* spore germination and outgrowth in ground turkey during chilling. J Food Prot 2007; 70(1): 218-22.
[http://dx.doi.org/10.4315/0362-028X-70.1.218] [PMID: 17265885]

[60]   Solórzano-Santos F, Miranda-Novales MG. Essential oils from aromatic herbs as antimicrobial agents. Curr Opin Biotechnol 2012; 23(2): 136-41.
[http://dx.doi.org/10.1016/j.copbio.2011.08.005] [PMID: 21903378]

[61]   Nourbakhsh F, Lotfalizadeh M, Badpeyma M, Shakeri A, Soheili V. From plants to antimicrobials: Natural products against bacterial membranes. Phytother Res 2022; 36(1): 33-52.
[http://dx.doi.org/10.1002/ptr.7275] [PMID: 34532918]

[62]   Lambert RJW, Skandamis PN, Coote PJ, Nychas GJE. A study of the minimum inhibitory concentration and mode of action of oregano essential oil, thymol and carvacrol. J Appl Microbiol 2001; 91(3): 453-62.
[http://dx.doi.org/10.1046/j.1365-2672.2001.01428.x] [PMID: 11556910]

[63] Yu C, Guo Y, Yang Z, Yang W, Jiang S. Effects of star anise (*Illicium verum* Hook.f.) essential oil on nutrient and energy utilization of laying hens. Anim Sci J 2019; 90(7): 880-6.
[http://dx.doi.org/10.1111/asj.13221] [PMID: 31111618]

[64] Park JH, Kang SN, Shin D, Shim KS. Antioxidant enzyme activity and meat quality of meat type ducks fed with dried oregano (*Origanum vulgare* L.) powder. Asian-Australas J Anim Sci 2015; 28(1): 79-85.
[http://dx.doi.org/10.5713/ajas.14.0313] [PMID: 25557678]

[65] Botsoglou NA, Florou-Paneri P, Christaki E, Fletouris DJ, Spais AB. Effect of dietary oregano essential oil on performance of chickens and on iron-induced lipid oxidation of breast, thigh and abdominal fat tissues. Br Poult Sci 2002; 43(2): 223-30.
[http://dx.doi.org/10.1080/00071660120121436] [PMID: 12047086]

[66] Luna A, Lábaque MC, Zygadlo JA, Marin RH. Effects of thymol and carvacrol feed supplementation on lipid oxidation in broiler meat. Poult Sci 2010; 89(2): 366-70.
[http://dx.doi.org/10.3382/ps.2009-00130] [PMID: 20075292]

[67] Mueller K, Blum NM, Kluge H, Mueller AS. Influence of broccoli extract and various essential oils on performance and expression of xenobiotic- and antioxidant enzymes in broiler chickens. Br J Nutr 2012; 108(4): 588-602.
[http://dx.doi.org/10.1017/S0007114511005873] [PMID: 22085616]

[68] Radwan Nadia L, Hassan RA, Qota EM, Fayek HM. Effect of natural antioxidant on oxidative stability of eggs and productive and reproductive performance of laying hens. Int J Poult Sci 2008; 7(2): 134-50.
[http://dx.doi.org/10.3923/ijps.2008.134.150]

[69] Liu XD, Jang A, Lee BD, Lee SK, Lee M, Jo C. Effect of dietary inclusion of medicinal herb extract mix in a poultry ration on the physico-chemical quality and oxidative stability of eggs. Asian-Australas J Anim Sci 2009; 22(3): 421-7.
[http://dx.doi.org/10.5713/ajas.2009.80334]

[70] Giannenas I, Florou-Paneri P, Papazahariadou M, Christaki E, Botsoglou NA, Spais AB. Effect of dietary supplementation with oregano essential oil on performance of broilers after experimental infection with *Eimeria tenella*. Arch Tierernahr 2003; 57(2): 99-106.
[http://dx.doi.org/10.1080/0003942031000107299] [PMID: 12866780]

[71] Oral N, Vatansever L, Sezer Ç, *et al.* Effect of absorbent pads containing oregano essential oil on the shelf life extension of overwrap packed chicken drumsticks stored at four degrees Celsius. Poult Sci 2009; 88(7): 1459-65.
[http://dx.doi.org/10.3382/ps.2008-00375] [PMID: 19531718]

[72] Gyamfi D, Awuah EO, Owusu S. Lipid metabolism: an overview. 2019.

[73] Semova I, Biddinger SB. Triglycerides in nonalcoholic fatty liver disease: guilty until proven innocent. Trends Pharmacol Sci 2021; 42(3): 183-90.
[http://dx.doi.org/10.1016/j.tips.2020.12.001] [PMID: 33468321]

[74] Chan J, Karere GM, Cox LA, VandeBerg JL. Animal models of diet-induced hypercholesterolemia. In: Kumar SA, Ed. Hypercholesterolemia. London: IntechOpen 2015.
[http://dx.doi.org/10.5772/59610]

[75] Ouimet M, Barrett TJ, Fisher EA. HDL and reverse cholesterol transport. Circ Res 2019; 124(10): 1505-18.
[http://dx.doi.org/10.1161/CIRCRESAHA.119.312617] [PMID: 31071007]

[76] Abo Ghanima MM, Alagawany M, Abd El-Hack ME, *et al.* Consequences of various housing systems and dietary supplementation of thymol, carvacrol, and euganol on performance, egg quality, blood chemistry, and antioxidant parameters. Poult Sci 2020; 99(9): 4384-97.
[http://dx.doi.org/10.1016/j.psj.2020.05.028] [PMID: 32867982]

[77]   Migliorini MJ, Boiago MM, Roza LF, *et al.* Oregano essential oil (*Origanum vulgare*) to feed laying hens and its effects on animal health. An Acad Bras Cienc 2019; 91(1): e20170901.
[http://dx.doi.org/10.1590/0001-3765201920170901] [PMID: 30785496]

[78]   Bolukbasi SC, Erhan MK. OZKAN A. Effect of dietary thyme oil and vitamin E on growth, lipid oxidation, meat fatty acid composition and serum lipoproteins of broilers. S Afr J Anim Sci 2006; 36(3): 189-96.

[79]   Sirvydis VH, Bobiniene R, Priudokiene V, Vencius D. Phytobiotics add value to broiler feed. Poult World 2003; 19(1): 16-7.

[80]   Hashemipour H, Kermanshahi H, Golian A, Veldkamp T. Effect of thymol and carvacrol feed supplementation on performance, antioxidant enzyme activities, fatty acid composition, digestive enzyme activities, and immune response in broiler chickens. Poult Sci 2013; 92(8): 2059-69.
[http://dx.doi.org/10.3382/ps.2012-02685] [PMID: 23873553]

[81]   Nitsan Z, Dunnington EA, Siegel PB. Organ growth and digestive enzyme levels to fifteen days of age in lines of chickens differing in body weight. Poult Sci 1991; 70(10): 2040-8.
[http://dx.doi.org/10.3382/ps.0702040] [PMID: 1720246]

[82]   Svihus B. Function of the digestive system. J Appl Poult Res 2014; 23(2): 306-14.
[http://dx.doi.org/10.3382/japr.2014-00937]

[83]   Ding X, Wu X, Zhang K, *et al.* Dietary supplement of essential oil from oregano affects growth performance, nutrient utilization, intestinal morphology and antioxidant ability in Pekin ducks. J Anim Physiol Anim Nutr (Berl) 2020; 104(4): 1067-74.
[http://dx.doi.org/10.1111/jpn.13311] [PMID: 31953905]

[84]   Gul M, Yilmaz E, Yildirim BA, *et al.* Effects of oregano essential oil (*Origanum syriacum* L.) on performance, egg quality, intestinal morphology and oxidative stress in laying hens. Eur Polit Sci 2019; 83: 290.
[http://dx.doi.org/10.1399/eps.2019.290]

[85]   Behnamifar A, Rahimi S, Torshizi MAK, Zade ZM. Effect of chamomile, wild mint, and oregano herbal extracts on quality and quantity of eggs, hatchability, and other parameters in laying Japanese quails. J. Med. Plants By-Product 2018; 7(2): 173-80.
[http://dx.doi.org/10.22092/jmpb.2018.118145]

[86]   Mohammadi Z, Ghazanfari S, Moradi MA. Effect of supplementing clove essential oil to the diet on microflora population, intestinal morphology, blood parameters and performance of broilers. Europ. Poult. Sci 2014; 78: eps.2014.51.
[http://dx.doi.org/10.1399/eps.2014.51]

[87]   Windisch W, Schedle K, Plitzner C, Kroismayr A. Use of phytogenic products as feed additives for swine and poultry1. J Anim Sci 2008; 86(14) (Suppl. 14): E140-8.
[http://dx.doi.org/10.2527/jas.2007-0459] [PMID: 18073277]

[88]   Alagawany M, Abd El-Hack ME, Eds. Natural feed additives used in the poultry industry. Sharjah: Bentham Science Publishers 2020.
[http://dx.doi.org/10.2174/97898114884501200101]

CHAPTER 9

Zingiber officinale (GINGER)

Youssef A. Attia[1,2,*], Nicola F. Addeo[3], Fulvia Bovera[3], Mohamed E. Abd El-Hack[4,*], Ayman E. Taha[5], Mohamed A. AlBanoby[6], Adel D. Al-qurashi[1], Asmaa F. Khafaga[7], Vincenzo Tufarelli[8], Mohamed W. Radwan[2] and Saber S. Hassan[2]

[1] *Sustainable Agriculture Production Research Group, Agriculture Department, , Faculty of Environmental Sciences, King Abdulaziz University, Jeddah-21589, Saudi Arabia*

[2] *Animal and Poultry Production Department, Faculty of Agriculture, Damanhour University, Damanhour-22713, Egypt*

[3] *Sustainable Agriculture Production Research Group, Department of Veterinary Medicine and Animal Production, University of Napoli Federico II, Via F. Delpino 1, 80137-Napoli, Italy*

[4] *Poultry Department, Faculty of Agriculture, Zagazig University, Zagazig-44519, Egypt*

[5] *Department of Animal Husbandry and Animal Wealth Development, Faculty of Veterinary Medicine, Alexandria University, Apis, Alexandria, 21944, Egypt*

[6] *Al-Shamel Animal Feed Factory, Industrial Area, Hail-55411, Saudi Arabia*

[7] *Department of Pathology, Faculty of Veterinary Medicine, Alexandria University, Apis, Alexandria, 21944, Egypt*

[8] *Department of Precision and Regenerative Medicine and Jonian Area (DiMePRe-J), Section of Veterinary Science and Animal Production, University of Bari 'Aldo Moro' s.p. Casamassima km 3, 70010 Valenzano, Italy*

Abstract: *Zingiber officinale*, commonly known as ginger, is a spicy plant with active ingredients such as gingerol and shogaol (Fig. **1**). Ginger has been widely used in traditional medicine to improve digestion, as it has been shown to increase the digestive enzyme protease (zingibain). Ginger is well known for its antibacterial and antiphlogistic properties, in addition to its ability to lower cholesterol levels in the bloodstream. In this chapter, we focus on the use of ginger as a feed supplement for enhancing poultry nutrition and the impact of this phenomenon on productive efficiency, carcass characteristics, hematology, gut microbiota, and toxicity.

Keywords: Blood biochemistry, Carcass traits, Gut microbiota, Ginger, Productive performance, Toxicity.

---

[*] **Corresponding authors Youssef A. Attia and Mohamed E. Abd El-Hack:** Sustainable Agriculture Production Research Group, Agriculture Department, Faculty of Environmental Sciences, King Abdulaziz University, Jeddah-21589, Saudi Arabia & Animal and Poultry Production Department, Faculty of Agriculture, Damanhour University, Damanhour-22713, Egypt and Poultry Department, Faculty of Agriculture, Zagazig University, Zagazig-44511, Egypt; E-mail: yaattia@kau.edu.sa

INTRODUCTION

Ginger, scientifically known as *Zingiber officinale*, exhibits properties that are both antioxidant and hypoglycemic in nature [1 - 5] along with providing immunological advantages and enhanced digestive functions. According to emerging evidence, ginger can also be utilized as an alternative phytogenic growth enhancer for antibiotics [6, 7]. Furthermore, some studies have found that ginger can have favorable effects on weight gain, feed efficiency, and survival rate in animals [8 - 11]. It has been seen that the recommended dose of ginger in the poultry diet should be around 1% [12 - 14]. However, an increase in the ginger dose by more than 1% would result in an increase in expenditure associated with nourishment [15, 16]. Moreover, the results from adding intermittent supplementation were similar to those obtained from continuous supplementation with ginger, and this reduced the cost with the same outcome, showing a better economic benefit [17 - 19]. Ginger was found to have a positive influence on improving the antimicrobial and antioxidant status of animals and this results in improving overall performance [2, 20, 21] and in enhancing the quality of poultry products [22, 23]. Antioxidants can mitigate the undesired effects of exposure to stressful conditions by modulating inflammatory responses and physiological functions. In addition, antioxidants can enhance the viability of animals by boosting immunity and product quality [24 - 26]. In addition, chickens receiving higher vitamin E supplementation than the commonly recommended levels could be a useful subject to enhance both performance and immunity [27]. The chemical composition and bioactive constituents of ginger are shown in Fig. (**1**).

Fig. (**1**). Ginger and its main constituents.

APPLICATION METHODS OF FEED ADDITIVES

In recent years, more attention has been paid to the application methods that feed additives, and this is a result of the low cost of their supplementation [9]. This may be most effective during stressful conditions (heat stress, vaccination, handling, transportation, diseases, and debeaking) and/or early ages of animals Nasir and Grashorn [18]. The importance of ginger application in broiler diets is illustrated in Fig. (**2**).

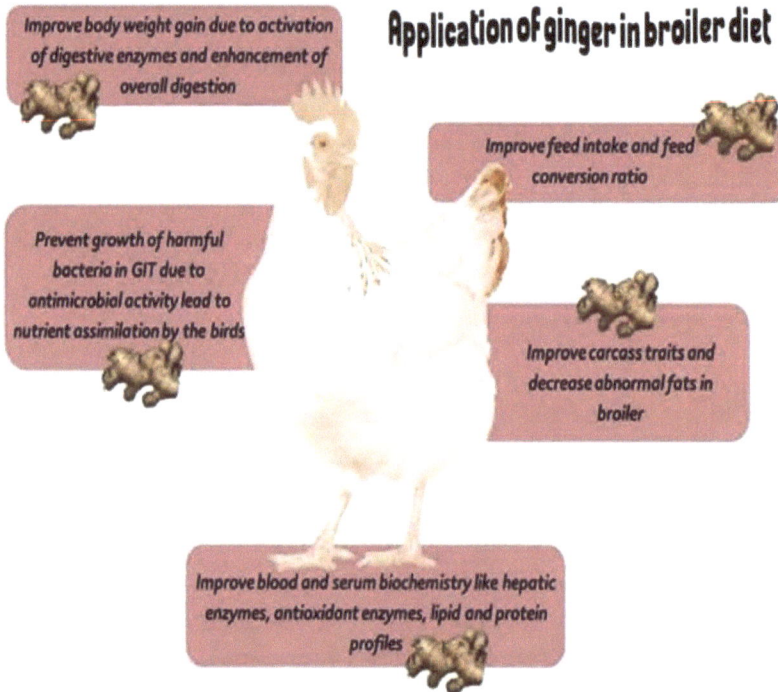

Application of ginger in broiler diet

- Improve body weight gain due to activation of digestive enzymes and enhancement of overall digestion
- Improve feed intake and feed conversion ratio
- Prevent growth of harmful bacteria in GIT due to antimicrobial activity lead to nutrient assimilation by the birds
- Improve carcass traits and decrease abnormal fats in broiler
- Improve blood and serum biochemistry like hepatic enzymes, antioxidant enzymes, lipid and protein profiles

Fig. (2). Importance of ginger application in broiler diet.

Growth and Body Weight Gain

In a study by Karangiya *et al.* [16], poultry diets supplemented with 1% ginger had a significantly higher feed intake (p< 0.05) than the untested group. In addition, the birds that were provided ginger-supplemented food experienced a higher rate of growth compared to the untested group. The supplemented broilers had a greater growth rate than the untested group, and their feed utilization was sustainably better (p<0.05). There was a marked increase in the intestinal villi width, length, and cryptal depth (p<0.05) in the ginger-fed group than in the other groups, indicating an enhanced absorptive surface area. There was no significant difference (p<0.05) in the cost of the feed and the European production index when compared to the untested group.

Habibi *et al.* [21] investigated how ginger affected accretion, carcass characteristics, antioxidant status, and hematology of broiler chick in conditions of thermal stress (32 ± 2°C for 8 h a day) using Cobb-500 broiler chickens. The chickens were fed a diet containing 100 mg/kg vitamin E, 7.5, or 15 g/kg ginger root powder (GRP), or 75 or 150 mg/kg ginger essential oil (GEO). At 3rd week of age, the group that received 7.5 g/kg of GPR group witnessed a significant boost in growth rate in contrast to the untested group. However, no notable disparities were observed across the different groups in relation to the growth rate, feed intake (FI), or feed utilization at 42 and 49 days of age.

George *et al.* [28] observed significant distinctions (p<0.05) in the consumption of feed, the increase in body weight, and the efficiency of feed transformation when varying concentrations of ginger (0%, 0.2%, 0.4%, and 0.6%) were administered. The results indicated that 0.6% ginger had the highest performance, indicating superior outcomes compared to the other levels (control, 0% ginger). Broilers fed up to 4 g GRP/kg diets showed increased growth performance [29]. Furthermore, Vencob-400 broilers fed diets containing 1% ginger powder showed a significantly increased body weight [30]. Meanwhile, ginger addition to broiler diets at 5-15 g/kg significantly lowered the bird growth rate compared to the untested group. Notable variations in accretion were observed among the experimental groups fed different ginger diets [31]. Dosu *et al.* [32] reported that up to 1.5% dietary ginger root extract (GRE) supplementation had no negative effects on growth performance measures compared to those of the untested group. Meanwhile, 3% GRE significantly decreased body weight gain (BWG) compared to the other groups.

Tekeli *et al.* [33] found that feeding *Z. officinale* from 120 to 360 ppm at 120 intervals to broiler chickens gradually increased growth; however, Zhang *et al.* [2] reported no changes in the growth rate of broilers when they were fed a diet containing 5 g/kg of ginger. In contrast, Herawati [34] observed that incorporating 2% *Alpinia purpurata* into the diet of broiler chickens resulted in increased body weight. Subsequently, Onu [35] discovered that supplementation of broiler diets with 0.25% ginger resulted in higher growth rates. In a study by Kausar *et al.* [36], the inclusion of ginger in a botanical solution at concentrations of 2 and 4 ml/l in the watering hole resulted in a significant impact on body weight at 35 days of the experiment. Similarly, Javed *et al.* [37] observed that the integration of ginger-infused broiler chicken feed resulted in improved weight gain.

Farinu *et al.* [38] revealed a slightly raised accretion of broiler chicks when integrated with 5, 10, or 15 g/kg of ginger. Conversely, Al-Homidan [15] discovered that when ginger was supplemented at 60 g/kg body weight, the growth rate of broilers was reduced from 1 to 4 weeks. Moorthy *et al.* [39] and

Zhang *et al.* [2] hypothesize that this could be a result of ginger's toxic effects. The potential reason for the inconsistent results observed in the experiments on broiler growth performance could be attributed to the different dosages employed.

Feed Consumption and Effectiveness

The available scientific literature presents conflicting opinions regarding the impact of ginger on feed intake. Rafiee *et al.* [40] highlighted that the use of 0.5% ginger extract, and thyme 0.5% extract increased feed intake in comparison to the control diet. Tekeli *et al.* [33] recorded a surge in feed consumption by broilers when 240 ppm *Z. officinale* was included in the diet. Nasiroleslami and Torki (2010) observed no impact on feed consumption or feed conversion ratio (FCR) in laying hens fed with the essential oil of ginger. Furthermore, according to Zhang *et al.* [2], the inclusion of ginger (5 g/kg) in diets of varying particle sizes (8.4, 37, 74, 149, and 300 µm) resulted in no notable variation in the daily feed consumption of broilers, although the feed intake of those fed ginger was marginally higher than that of the untested group. Zhao *et al.* [41] noticed no variation in feed consumption and FCR among laying hens provided with ginger (5, 10, 15 and 20 g/kg). Similarly, Akbarian *et al.* [42] observed no significant impact on feed consumption and FCR in laying hens fed ginger at distinct levels (0.25, 0.5, and 0.75%). Birds fed diets fortified with ginger levels of 0, 0.2, 0.4, and 0.6% displayed an improved FCR and reduced feed intake compared to the untested group [31]. Shewita and Taha [29] reported that broilers supplemented with up to 4 g ginger powder/kg feed improved FCR in broilers.

Meanwhile, Al-Khalaifah *et al.* [43] found that ginger supplementation in broiler diets at levels ranging from to 5-15 g/kg diet substantially reduced feed intake in relation to the untested group. Recently, Dosu *et al.* [32] reported that up to 1.5% dietary supplementation with ginger root extract (GRE) had no damaging effects on growth performance compared to that of the untested group. Meanwhile, 3% GRE significantly reduced the FCR compared to the other groups. In addition, Incharoen and Yamauchi [44] discovered that incorporating 1 and 5% of fermented dried ginger into the diet of White Leghorn laying hens led to an increase in feed consumption and FCR within the groups fed with ginger. Similarly, the inclusion of 2% *Alpinia purpurata* in broiler chicken diets led to enhanced feed consumption and FCR [34]. Additionally, Onu [35] discovered that including ginger (0.25%) in the control diet improved the FCR of broilers without changing the feed intake.

Javed *et al.* [37] demonstrated that the addition of an extract of a blend of botanicals comprising ginger can be used as a supplementary substance of broiler chickens at a concentration of 15 ml/l, which entails enhanced feed consumption and improved FCR. In addition, Kausar *et al.* [36] discovered that the administration of a botanical solution containing ginger at concentrations of 2 and 4 ml/l in water fed to broilers resulted in a higher FCR.

Moorthy *et al.* [39] found that there was no variation in feed consumption at 42 days of age in chickens that were given mixed species botanical products, including ginger, although a significantly higher FCR resulted in chickens fed a mixture of 0.2% ginger and 0.2% curry leaf powder. The disparities observed in the aforementioned results could be attributed to the variability in the types of ginger used, how they were processed, dosage, and the period of the trials. The majority of scientists ascribe the improvement in the performance of chickens fed ginger to improve appetite and enhanced digestion when used at low doses [16, 28 - 30, 32, 33, 35 - 37, 40].

They further hypothesized that because of this natural product, the flow of the feed in the digestive tract would have been faster, resulting in a faster emptied digestive tract and higher feed consumption. Ginger enhances the production of various intestinal enzymes, including lipase, disaccharidase, and maltase [2]. The enhanced productive performance of chickens that received ginger supplementation can be ascribed to the presence of two intestinal enzymes, protease, and lipase, present in ginger, which belong to the innate immunity of the plant [2, 34]. Ginger improves animal digestion and absorption of nutrients owing to its advantageous action on digestive activities, gastric secretion, and enterokinase [41].

Carcass Traits

Rafiee *et al.* [40] showed that the liver percentage of chickens fed with 0.5% ginger was significantly decreased. There were no notable variations in the heart percentage between the groups under trial. Moreover, the use of 0.5% ginger extract significantly reduced the abdominal fat percentage. Habibi *et al.* [21] checked out the influence that ginger feed supplementation has on carcass characteristics of heat-stressed ($32 \pm 2°C$ for 8 h per d) Cobb-500 broiler chickens. They pointed out that the inclusion of GRP at 7.5 and 15 g/kg and (GEO) at 75 and 150 mg/kg did not affect the carcass characteristics and inner organs of broiler chickens.

Saranya *et al.* [23] reported that tenderizer and the tenderization of meat of the ginger extract-fed broilers is obtained by ginger action on the constituents of toughness of both connective tissue and myofibrillar. Ginger used for tenderizing tough meat does not pose safety or aesthetic concerns among consumers [23]. Thus, ginger extract could enrich tough meat quality and prove to be profitable to the meat industry due to its proteolytic activity on collagen.

In addition, ginger extract displayed proteolytic activity, showing an increase in solubility and proteolysis in the collagen of the ginger extract-treated hen muscle [22]. Ginger stimulates salivary secretion [45]. It enhances the release of bile [46]. Ginger is utilized as an excitant and carminative for dyspepsia and colic [45]. Shewita and Taha [29] fed broiler chicks with ginger powder at levels of 2, 4, and 6 g/kg diet and found non-significant differences for most of the carcass traits except a significant decrease in abdominal fat at 4 and 6 g/ kg diet compared to other groups. Qorbanpour *et al.* [47] supplemented broiler chicks (Ross-308) with various levels of ginger powder (0.15, 0.20, and 0.25%) and did not detect notable variations among the interventions for most carcass traits except a marked diminution in the weight of gizzard and abdominal fat confronted with the untested group. Moreover, Egenuka *et al.* [48] supplemented diets of broilers with a ginger meal at concentrations of 0.0, 0.5, 1.0%, and 1.5% and reported a significant decrease in abdominal fat at the 1.5% level compared to other treated groups. In addition, El-Kashef [49] supplemented the diet of broiler chickens with ginger at 0.25, 0.50%, and 0.75% and reported a significant increase not only in carcass yield but also in thyme percentage, bursa, and spleen compared to untreated chickens. However, processing losses showed minor changes [50].

Zhang *et al.* [2] observed that the addition of ginger to birds' diets led to a greater carcass weight confronted with the untested group. Furthermore, ginger supplementation significantly improves breast and leg weight [37]. According to Zhang *et al.* [2], there is an association between the improvement in chicken carcass quality and the antioxidant effect of ginger, which has a positive effect on protein and fat metabolism. Paradoxically, Moorthy *et al.* [39] discovered that there was no percentage change in carcass weight, viscera, abdominal fat, or offal when ginger was administered. El-Deek *et al.* [51] reported that there was no difference among broilers that were subjected to control conditions and those treated with ginger up to 42 days of age. Similarly, Onu [35] found no notable changes in carcass traits when broiler chicks were fed a basal diet supplemented with ginger (0.25%). The remarkable performance of chickens could be attributed to antioxidant, antimicrobial, and other activities, such as elevated blood flow and digestive enzyme secretion, as well as low feed oxidation in ginger-fed animals [41].

Antioxidant Status

Free radicals are typically generated within the body, leading to oxidative harm inflicted on cells. Consequently, the presence of neutralizing agents is imperative for mitigating peroxidative injury. The body's defense mechanism against oxidation encompasses a combination of natural and synthetic antioxidants as well as antioxidant enzymes [2, 52]. Although free radicals are continuously produced within the body and play a crucial role in maintaining normal body functions, excessive production beyond the physiological range can have detrimental effects on cell membranes and organelles [27]. Three major antioxidant enzymes are present in the body: superoxide dismutase (SOD), glutathione peroxidase (GSHPx), and catalase, which work together to combat reactive oxygen species (ROS) [2, 41]. Always about ROS, current research by Addeo *et al.* (2021), demonstrated that incorporating vegetable-origin products into the diet of *Hermetia illucens* at varying levels has led to a notable decrease in the production of reactive oxygen species. Therefore, a higher level of this antioxidant can enhance the ability of the organism to scavenge and neutralize these harmful particles. Malondialdehyde (MDA) is among the most important aldehydes produced during the secondary oxidation of polyunsaturated fatty acids; its measurement serves as a key indicator of lipid peroxidation occurring inside the cell [53]. Moreover, Toghyani *et al.* [54] concluded that ginger significantly enhances the serum antioxidant capacity of broiler chicks but cannot be used as an immunomodulator. In addition, chicken diets at 0, 5, 10, or 15 g/kg showed no changes in liver T-SOD in broiler chicks fed a diet enriched with 5 g/kg of ginger. Nevertheless, the broilers that were fed a diet containing either 10 or 15 g/kg of ginger exhibited a greater level of liver T-SOD when compared to the groups that were fed 0 and 5 g/kg of ginger [43]. The same authors reported that liver GPX levels were unaffected by feeding regimens that included ginger supplements. Ginger addition raised CAT levels in front of the untested group. In contrast, the liver MDA of broiler chicks that were fed varying amounts of ginger was lower than that of the untested group.

Shah *et al.* [55] examined broilers exposed to heat stress with dietary ginger and onion levels as 5-15g/kg ginger and 1.5-3.5 g/kg onion; the results revealed that supplementing with onions and ginger significantly raised PON1 values and lowered MDA levels. Paraoxonase neutralizes free oxygen radicals; however, as the temperature increases over the thermoneutral zone, its concentration decreases. Namdeo *et al.* [50], reported that ginger-treated broilers showed significantly higher TBARS in liver and breast muscle tissue. Similarly, Sahoo *et al.* [56] reported a reduction in serum MDA levels in ginger-supplemented broiler diets at levels ranging from 0.5-1%.

Habibi *et al.* [21] researched the influence of ginger supplementation on the carcass characteristics of chickens exposed to heat stress ($32 \pm 2°C$ for 8 h per day) using Cobb-500 broilers. The results of their study demonstrated that the administration of 150 mg/kg GEO resulted in an elevation of superoxide dismutase (SOD) activity in the liver when confronted with the untested group. Additionally, the groups receiving GRP and GEO exhibited reduced levels of MDA in the liver compared to the untested group. However, there were no notable variations in the levels of SOD, catalase (CAT), or glutathione peroxidase enzymes in blood cells among the different groups. The same authors added that all dietary groups had an increasing effect on total antioxidant capacity (TAC) and reduced MDA levels in the serum compared to the untested group. The findings of this study indicate that GRP and GEO can potentially be used as substitutes for synthetic antioxidants in broiler chickens. Moreover, it appears that GRP might be better than GEO at enhancing the antioxidant effects of broilers.

Zhang *et al.* [2] found that supplementing with ginger at 5 g/kg had the marked effect of increasing GSHPx and SOD while lowering MDA levels at 21 and 42 days of age in broilers. This decreased MDA level indicated that ginger was capable of mitigating lipid peroxidation damage in cells. These authors suggested that the antioxidant effect of ginger was possibly due to a change in the availability of the active compounds within the ginger because the size was reduced from 300 to 37 μm. However, it was also noted that a particle size reduction (8 μm) could negatively affect the antioxidant status owing to the mechanical harm and oxidation of the active compounds that would occur when exposed to ambient air. Consequently, the efficacy of beneficial compounds in ginger may be diminished.

During the laying period, the integration of ginger into the diet of hens resulted in a favorable impact on the enzymatic antioxidant activity found in both the serum and egg yolk. This positive effect was observed at both 5 and 10 weeks of laying period [41]. Moreover, it has been found that adding ginger to diets at rates of 0.5 and 0.75% caused a significant increase in GSHPx and a reduction in MDA levels in plasma [42]. The potential explanation for elevated levels of antioxidant enzymes and the decrease in MDA levels could be attributed to the presence of ginger, although the precise mechanism remains unexplained.

Several studies have shown that antioxidant compounds are abundant in plant polyphenolic flavonoids [57 - 62]. Ginger contains various active components, including gingerol, gingerdiol, gingerdione shogaoli, and phenolic ketone products that show antioxidant activities [5, 41]. Research has indicated that both raw ginger plant materials and individual components such as gingerol may offer protection against lipid peroxidation [63, 64]. They also observed that the increase in serum antioxidants might be partly attributed to the delayed oxidation activity of the feed by GRP.

Blood Biochemistry

Rafiee *et al.* [40] showed that the triglyceride level was lower in chicks when 0.5% ginger was used, and the cholesterol level was at the lowest. Calcium and Phosphorus levels were higher in chicks fed a 0.5% ginger diet (p<0.05). At day

42, the antibody titer against Newcastle disease virus (NDV) showed that it was significantly higher in the 0.5% ginger extract group (p<0.05). Habibi *et al.* [21] showed the effect of ginger on carcass properties of heat-stressed ($32 \pm 2°C$ for 8 h per d) Cobb-500 broiler chickens; the results showed there was no influence on blood parameters like lipid profiles, glucose, total protein, and albumen when GRP (7.5 and 15g/kg) and GEO (75 and 150mg/kg) in broiler diets were included.

Ginger supplementation at the levels of 2, 4, and 6 g/kg diets was reported by Shewiata and Taha [29] to significantly increase broiler chicks' immunity against NDV, as indicated by an increase in hemagglutination inhibition (HI) titer. Additionally, a broiler diet enriched with ginger at a dose of 6 g/kg showed a substantial increase in white blood cells and a significant decrease in HDL and cholesterol levels in the same group. Furthermore, there was a substantial decrease in both VLDL and TG levels between the groups supplemented with ginger and the untested group. Saeid *et al.* [65] discovered a considerable reduction in serum glucose, low-density lipoprotein (LDL-cholesterol), very low-density lipoprotein (vLDL-cholesterol), and total cholesterol, levels among the groups that received ginger extract at concentrations of 0.4% and 0.6%. Interestingly, the chickens in these groups exhibited elevated high-density lipoprotein (HDL-cholesterol) cholesterol concentrations.

Onu [35] conducted a study that investigated the effects of ginger integration in broiler chicks. The study revealed that adding 0.25% ginger to the basal diet did not result in any significant changes in the urea, creatine, total protein, globulin, and albumin levels. Supplementing broilers with 10 ml/L of a combination of berberine, garlic, Aloe vera, and ginger in their drinking water resulted in a notable reduction in serum glucose, alanine aminotransferase (ALT), aspartate aminotransferase (AST), and alkaline phosphatase (ALP) concentrations in the treated group. In the same experiment, the cholesterol profiles of the broilers were assessed. The treated group exhibited a notable decrease in the total cholesterol, triglyceride, LDL, and vLDL concentrations. Conversely, HDL cholesterol levels increased.

Zhang *et al.* [2] focused on the effects of ginger powder on broilers at 3 and 6 weeks. The findings indicated that the total protein concentration increased, whereas the cholesterol levels decreased in broilers treated with ginger powder. Kausar *et al.* [36] found that there was no effect on total protein, globulin and serum albumin in broilers who received diets containing a carminative mixture with ginger at 2 and 4 ml/L in a watering hole. Al-Homidan's [15] study, the inclusion of 60 g/kg of ginger in the diet of broiler chicks resulted in a reduction in globulin levels and plasma total protein. This decrease could be attributed to the harmful effects of excess ginger dose.

Conversely, Farinu *et al.* [38] reported no significant effect on serum total protein and albumin levels in broiler chickens fed ginger at doses ranging from 5 to 15 g/kg diet. The variations in these findings could be attributed to differences in the administered doses or variations in experimental settings. The exact mechanisms underlying alterations in blood metabolites remain elusive. Nevertheless, it has been postulated that (E)-8 beta, 17-epoxyllabed-12-ene-15, 16-dial, a ginger-derived compound, may impede cholesterol biosynthesis in the liver of hypercholesterolemic mice, thereby leading to a reduction in cholesterol levels [66]. Srinivasan and Sambaiah [67] demonstrated that when rats were fed ginger, there was a substantial rise of the hepatic cholesterol enzyme 7-alpha-hydroxylase, which is the enzyme least present in the formation of bile acids. Its presence aids in the excretion of cholesterol from the body.

Gut Function

Extensive research has been conducted on the reaction of the gastrointestinal tract in broilers to nutrient intake [68]. Dietary feed components commonly affect the histology of the intestinal apical surface [69, 70]. Moreover, it has been suggested that a greater villus height can lead to a higher absorption of accessible nutrients because of the expanded surface area [29, 71, 72]. Greater villus height and a higher rate of mitosis in the gut are indicators of increased stimulation of intestinal functions as a consequence of elevated nutrient absorption [73 - 75]. Increased villous height in the intestine is an indicator of the stimulation of intestinal function due to high nutrient absorption [76].

Incharoen and Yamauchi [44] observed that the histological parameters of the intestinal segments in laying hens fed a ginger-containing diet were significantly greater than those in control hens. Furthermore, the number of filamentous bacteria was found to be increased, which possess immunomodulating properties evidenced by higher levels of immunoglobulin A (IgA) in the intestine. Garland *et al.* [77] and Heczko *et al.* [78] deemed it appropriate to highlight how this leads to protection against *Salmonella enteritidis* and *Escherichia coli*. Shewiata and Taha [29] broiler chicks supplemented with ginger at levels of 2, 4, and 6 g/kg diets and found that the ginger-supplemented groups showed greater crypt depths and higher villus lengths compared to the untested group.

Kausar *et al.* [36] noticed that a botanical solution containing ginger at 4 ml/l of drinking water caused an increase in means to titer in responses against Newcastle disease, stating that ginger has an immunomodulating effect. Sudrashan *et al.* [79] demonstrated how the essential oil isolated from ginger and added as a disinfectant in different concentrations to chicken meat showed a significant decrease in *Staphylococcus*, *E. coli* and *Salmonella spp*. Zhao *et al.* [41] observed that the inclusion of ginger in laying hen feed resulted in decreased oxidation, which could be attributed to improved laying performance and increased antioxidant levels in both the whey and yolk of the chicken.

Khan *et al.* [20] carried out a review of published research up to 2012 and concluded that ginger had varying levels of efficacy in chickens' diets; there were signs suggesting that the addition of this botanical substance in the diet of chickens could enhance their growth performance, while also promoting improved gastrointestinal function and antioxidant activity in poultry. Nevertheless, the need for standardization of doses, application (via feed or water), and extraction processes was highlighted in order to draw firm conclusions regarding its efficacy. Subsequent studies should focus on this standardization to maximize the benefits of ginger, whether used in feed or water, for poultry producers.

Toxicological Impacts

The toxicity of dietary supplements may vary according to the amount and duration of administration [34]. Research on the potential toxic effects of incorporating ginger into poultry feed remains scarce. However, Herawati [34] documented those broilers fed diets containing 0.5%, 1.0%, and 1.5% *Alpinia purpurata* experienced adverse effects such as muscle edema, necrosis, and inflammation. This is because of the toxic sesquiterpene oil present in ginger, which has a toxic potential in animals. High doses of these substances can cause congestion, edema, inflammation, and necrosis [34, 80].

CONCLUSION AND REMARKS

Ginger has shown promising potential as a feed additive to enhance poultry nutrition and performance. When used at appropriate levels, ginger supplementation can improve production rate, feed efficiency, carcass quality, antioxidant status, and gut health in poultry. The active compounds in ginger, such as gingerols and shogaols, appear to have beneficial effects on digestion, metabolism, and immune functions. However, the optimal dosage is critical because excessive amounts may have adverse effects. While most studies report positive outcomes with ginger supplementation up to 1-2% of the diet, some studies indicate potential toxicity at higher levels, particularly with certain varieties, such as red ginger. Overall, ginger is a natural feed additive that can enhance the productive efficiency and product quality of poultry. Further research is needed to standardize application methods and determine ideal inclusion rates for different poultry production systems and organic poultry production as natural products.

IMPLICATIONS

- Ginger fortification of poultry diets has significant implications for the poultry industry and animal nutrition research. The potential use of ginger as a natural feed additive could lead to improved productivity, product quality, and animal welfare in commercial poultry production.
- The variation across studies highlights the need for standardized protocols and further research to determine the optimal dosage, application methods, and ginger varieties for different poultry production systems.
- The antioxidant and antimicrobial properties of ginger may also have implications for reducing the use of synthetic additives and antibiotics in poultry feed, aligning with the growing consumer demand for natural and antibiotic-free poultry products.
- The potential toxicity of higher (at dose, 5-15 g/kg diets) showed some toxic effects that warrant further investigation, emphasizing the importance of careful dosage control and toxicological studies to ensure safe implementation.
- Ginger fortification could be a valuable tool for enhancing poultry nutrition and performance; however, its effective and safe application warrants further research and development for animal nutrition.
- Ginger root powder up to 7.5 g/kg diet and ginger essential oil up to 75 mg/kg feed are valuable feed supplements for poultry. They can enhance growth performance, carcass traits, gut microbiota composition, and antioxidant status.

REFERENCES

[1] Al-Amin ZM, Thomson M, Al-Qattan KK, Peltonen-Shalaby R, Ali M. Anti-diabetic and hypolipidaemic properties of ginger (*Zingiber officinale*) in streptozotocin-induced diabetic rats. Br J Nutr 2006; 96(4): 660-6.
[http://dx.doi.org/10.1079/BJN20061849] [PMID: 17010224]

[2] Zhang GF, Yang ZB, Wang Y, Yang WR, Jiang SZ, Gai GS. Effects of ginger root (*Zingiber officinale*) processed to different particle sizes on growth performance, antioxidant status, and serum metabolites of broiler chickens. Poult Sci 2009; 88(10): 2159-66.
[http://dx.doi.org/10.3382/ps.2009-00165] [PMID: 19762870]

[3] Morakinyo AO, Akindele AJ, Ahmed Z. Modulation of antioxidant enzymes and inflammatory cytokines: Possible mechanism of anti-diabetic effect of ginger extracts. Afr J Biomed Res 2011; 9: 195-202.

[4] Khafaga AF, Abd El-Hack ME, Taha AE, Elnesr SS, Alagawany M. The potential modulatory role of herbal additives against Cd toxicity in human, animal, and poultry: a review. Environ Sci Pollut Res Int 2019; 26(5): 4588-604.
[http://dx.doi.org/10.1007/s11356-018-4037-0] [PMID: 30612355]

[5] Abd El-Hack ME, Alagawany M, Shaheen H, *et al.* Ginger and its derivatives as promising alternatives to antibiotics in poultry feed. Animals (Basel) 2020; 10(3): 452.
[http://dx.doi.org/10.3390/ani10030452] [PMID: 32182754]

[6] Demir E, Sarica S, Ozcan MA, Suicmez M. The use of natural feed additives as alternatives for an antibiotic growth promoter in broiler diets. Br J Polit Sci 2003; 44: S44-5.

[7] Abd El-Hack ME, Abdelnour SA, Taha AE, *et al.* Herbs as thermoregulatory agents in poultry: An

overview. Sci Total Environ 2020; 703: 134399.
[http://dx.doi.org/10.1016/j.scitotenv.2019.134399] [PMID: 31757531]

[8]   Issa KJ, Omar JMA. Effect of garlic powder on performance and lipid profile of broilers. Open J Anim Sci 2012; 2(2): 62-8.
[http://dx.doi.org/10.4236/ojas.2012.22010]

[9]   Oleforuh-Okoleh VU, Chukwu GC, Adeolu AI. Effect of ground ginger and garlic on the growth performance, carcass quality and economics of production of broiler chickens. Glob J Biosci Biotechnol 2014; 3(3): 225-9.

[10]   Amber K, Badawy NA, El-Sayd AENA, Morsy WA, Hassan AM, Dawood MAO. Ginger root powder enhanced the growth productivity, digestibility, and antioxidative capacity to cope with the impacts of heat stress in rabbits. J Therm Biol 2021; 100: 103075.
[http://dx.doi.org/10.1016/j.jtherbio.2021.103075] [PMID: 34503812]

[11]   Cardoso AJS, dos Santos WV, Gomes JR, *et al.* Ginger oil, *Zingiber officinale*, improve palatability, growth and nutrient utilisation efficiency in Nile tilapia fed with excess of starch. Anim Feed Sci Technol 2021; 272: 114756.
[http://dx.doi.org/10.1016/j.anifeedsci.2020.114756]

[12]   Eltazi MA. Response of broiler chicks to diets containing different mixture levels of garlic and ginger powder as natural feed additives. Int J Pharm Res Allied Sci 2014; 3(4): 27-35.

[13]   Bamidele O, Adejumo IO. Effect of garlic (*Allium sativum* L.) and ginger (*Zingiber officinale* Roscoe) mixtures on performance characteristics and cholesterol profile of growing pullets. Int J Poult Sci 2012; 11(3): 217-20.
[http://dx.doi.org/10.3923/ijps.2012.217.220]

[14]   Al-Khalaifah H. Al-NasserA, Al-Surrayai T, Sultan H, Al-Attal D, Al-Kandari R, Al-Saleem H, Al-Holi A, Dashti F. Effect of ginger powder on production performance, antioxidant status, haematological parameters, digestibility, and plasma cholesterol content in broiler chickens. Animals (Basel) 2022; 12(7): 901.
[http://dx.doi.org/10.3390/ani12070901] [PMID: 35405889]

[15]   Al-Homidan AA. Efficacy of using different sources and levels of *Allium sativum* and *Zingiber officinale* on broiler chicks performance. Saudi J. of Biological Sci 2005; s 12: 96–102.

[16]   Karangiya VK, Savsani HH, Patil SS, *et al.* Effect of dietary supplementation of garlic, ginger and their combination on feed intake, growth performance and economics in commercial broilers. Vet World 2016; 9(3): 245-50.
[http://dx.doi.org/10.14202/vetworld.2016.245-250] [PMID: 27057106]

[17]   Grashorn M. Use of phytobiotics in broiler nutrition – an alternative to infeed antibiotics? J Anim Feed Sci 2010; 19(3): 338-47.
[http://dx.doi.org/10.22358/jafs/66297/2010]

[18]   Nasir Z, Grashorn MA. Effects of intermittent application of different *Echinacea purpurea* juices on broiler performance and some blood parameters. Arch Geflugelkd 2010; 74: 36-42.

[19]   Attia YA, Al-Hamid AEA, Ibrahim MS, Al-Harthi MA, Bovera F, Elnaggar AS. Productive performance, biochemical and hematological traits of broiler chickens supplemented with propolis, bee pollen, and mannan oligosaccharides continuously or intermittently. Livest Sci 2014; 164: 87-95.
[http://dx.doi.org/10.1016/j.livsci.2014.03.005]

[20]   Khan RU, Naz S, Nikousefat Z, *et al.* Potential applications of ginger (*Zingiber officinale*) in poultry diets. Worlds Poult Sci J 2012; 68(2): 245-52.
[http://dx.doi.org/10.1017/S004393391200030X]

[21]   Habibi R, Sadeghi GH, Karimi A. Effect of different concentrations of ginger root powder and its essential oil on growth performance, serum metabolites and antioxidant status in broiler chicks under heat stress. Br Poult Sci 2014; 55(2): 228-37.

[http://dx.doi.org/10.1080/00071668.2014.887830] [PMID: 24697550]

[22]   Naveena BM, Mendiratta SK. Tenderisation of spent hen meat using ginger extract. Br Poult Sci 2001; 42(3): 344-9.
[http://dx.doi.org/10.1080/00071660120055313] [PMID: 11469554]

[23]   Saranya S, Santhi D, Kalaikannan A. Ginger as a tenderizing agent for tough meats- A review. J Livest Sci 2016; 7: 54-61.

[24]   Attia YA, Abd-El-Hamid AE, Abd El-Ghany FA, Habiba HI. Effect of oil source and antioxidant supplementations on growth performance and meat quality of Japanese quail male. Proceeding of XII European Poult. Conference, Verona, Italy, 10-14 Sept Abstract in. Worlds Poult Sci J 2006; 62 (Suppl.): 405.

[25]   Dalólio FS, Albino LFT, Lima HJD, Silva JN, Moreira J. Heat stress and vitamin E in diets for broilers as a mitigating measure. Acta Sci Anim Sci 2015; 37(4): 419-27.
[http://dx.doi.org/10.4025/actascianimsci.v37i4.27456]

[26]   Attia YA, Abd El-Hamid AEHE, Abedalla AA, *et al.* Laying performance, digestibility and plasma hormones in laying hens exposed to chronic heat stress as affected by betaine, vitamin C, and/or vitamin E supplementation. Springerplus 2016; 5(1): 1619.
[http://dx.doi.org/10.1186/s40064-016-3304-0] [PMID: 27652192]

[27]   Khan RU. Antioxidants and poultry semen quality. Worlds Poult Sci J 2011; 67(2): 297-308.
[http://dx.doi.org/10.1017/S0043933911000316]

[28]   George OS, Kaegon SG, Igbokwe AA. Effects of graded levels of ginger (*Zingiber officinale*) meal as feed additive on growth performance characteristics of broiler chicks. Inter. J. of Sci. and Res 2013; 4: 805-808.

[29]   Shewita RS, Taha AE. Influence of dietary supplementation of ginger powder at different levels on growth performance, haematological profiles, slaughter traits and gut morphometry of broiler chickens. S Afr J Anim Sci 2018; 48(6): 997-1008.

[30]   Gaikwad DS, Fulpagare YG, Bhoite UY, Deokar DK, Nimbalkar CA. Effect of dietary supplementation of ginger and cinnamon on growth performance and economics of broiler production. Int J Curr Microbiol Appl Sci 2019; 8(3): 1849-57.
[http://dx.doi.org/10.20546/ijcmas.2019.803.219]

[31]   Kairalla MA, Aburas AA, Alshelmani MI. Effect of diet supplemented with graded levels of ginger (*Zingiber officinale*) powder on growth performance, haematological parameters, and serum lipids of broiler chickens. Arch Razi Inst 2022; 77(6): 2089-95.
[PMID: 37274916]

[32]   Dosu G, Obanla TO, Zhang S, *et al.* Supplementation of ginger root extract into broiler chicken diet: effects on growth performance and immunocompetence. Poult Sci 2023; 102(10): 102897.
[http://dx.doi.org/10.1016/j.psj.2023.102897] [PMID: 37562125]

[33]   Tekeli A, Kutlu HR, Celik L. Effects of *Z. officinale* and propolis extracts on the performance, carcass and some blood parameters of broiler chicks. Current Research in Poultry Science 2010; 1(1): 12-23.
[http://dx.doi.org/10.3923/crpsaj.2011.12.23]

[34]   Herawati . The effect of feeding red ginger as phytobiotic on body weight gain, feed conversion and internal organs condition of broiler. Int J Poult Sci 2010; 9(10): 963-7.
[http://dx.doi.org/10.3923/ijps.2010.963.967]

[35]   Onu PN. Evaluation of two herbal spices as feed additives for finisher broilers. Biotechnol Anim Husb 2010; 26(5-6): 383-92.
[http://dx.doi.org/10.2298/BAH1006383O]

[36]   Kausar R, Rizvi F, Anjum AD. Effect of carminative mixture on health of broiler chicks. Pakistan J. of Biological Sci 1999; s 2: 1074-1077.
[http://dx.doi.org/10.3923/pjbs.1999.1074.1077]

[37]   Javed M, Durrani F, Hafeez A, Khan RU, Ahmed I. Effect of aqueous extract of plant mixture on carcass quality of broiler chicks. J Agric Biol Sci 2009; 4: 37-40.

[38]   Farinu GO, Ademola SG, Ajayi AO, Babatunde GM. Growth, haematological and biochemical studies on garlic and ginger-fed broiler chickens. Moor. J Agric Res (Lahore) 2004; 5: 122-8.

[39]   Moorthy M, Ravi S, Ravikuma M, Viswanatha K, Edwin SC. Ginger, Pepper and Curry Leaf Powder as Feed Additives in Broiler Diet. Int J Poult Sci 2009; 8(8): 779-82.
[http://dx.doi.org/10.3923/ijps.2009.779.782]

[40]   Rafiee A, Rahimian Y, Zamani F, Asgarian F. Effect of use ginger (*Zingiber officinale*) and thymus (*Thymus vulgaris*) extract on performance and some haematological parameters on broiler chicks. Sci Agric 2013; 4: 20-5.

[41]   Zhao X, Yang ZB, Yang WR, Wang Y, Jiang SZ, Zhang GG. Effects of ginger root (*Zingiber officinale*) on laying performance and antioxidant status of laying hens and on dietary oxidation stability. Poult Sci 2011; 90(8): 1720-7.
[http://dx.doi.org/10.3382/ps.2010-01280] [PMID: 21753209]

[42]   Akbarian A, Golian A, Sheikh Ahmadi A, Moravej H. Effects of ginger root (*Zingiber officinale*) on egg yolk cholesterol, antioxidant status and performance of laying hens. J Appl Anim Res 2011; 39(1): 19-21.
[http://dx.doi.org/10.1080/09712119.2011.558612]

[43]   Al-Khalaifah H, Al-Nasser A. Al-SurrayaiT, Sultan H, Al-Attal D, Al-Kandari R, Al-Saleem H, Al-Holi A, Dashti F. Effect of ginger powder on production performance, antioxidant status, haematological parameters, digestibility, and plasma cholesterol content in broiler chickens. Animals (Basel) 2022; 12(7): 901.
[http://dx.doi.org/10.3390/ani12070901] [PMID: 35405889]

[44]   Incharoen T, Yamauchi K. Production performance, egg quality and intestinal histology in laying hens fed dietary dried fermented ginger. Int J Poult Sci 2009; 8(11): 1078-85.
[http://dx.doi.org/10.3923/ijps.2009.1078.1085]

[45]   O'Hara M, Kiefer D, Farrell K, Kemper K. A review of 12 commonly used medicinal herbs. Arch Fam Med 1998; 7(6): 523-36.
[http://dx.doi.org/10.1001/archfami.7.6.523] [PMID: 9821826]

[46]   Kato M, Rocha MLR, Carvalho AB, Chaves MEC, Raña MCM, Oliveira FC. Occupational exposure to neurotoxicants: preliminary survey in five industries of the Camaçari Petrochemical Complex, Brazil. Environ Res 1993; 61(1): 133-9.
[http://dx.doi.org/10.1006/enrs.1993.1057] [PMID: 8472667]

[47]   Qorbanpour M, Fahim T, Javandel F, *et al.* Effect of dietary ginger (*Zingiber officinale* Roscoe) and multi-strain probiotic on growth and carcass traits, blood biochemistry, immune responses and intestinal microflora in broiler chickens. Animals (Basel) 2018; 8(7): 117.
[http://dx.doi.org/10.3390/ani8070117] [PMID: 30011890]

[48]   Egenuka FC, Achi JC, Obi HN, Okere PC, Kadurumba OE, Iwuji TC. Effect of different inclusion levels of ginger (*Zingiber officinale* Roscoe) meal on performance, cost implication and carcass characteristics of broiler chickens. Int J Agric Rural Dev 2021; 24(2): 5923-9.

[49]   El-Kashef M. Evaluation of using ginger (*Zingiber officinale*) on growth performance, carcass characteristics, blood biochemistry and immune responses of quail birds. Egypt Poult Sci 2022; 42(2): 199-212.
[http://dx.doi.org/10.21608/epsj.2022.249546]

[50]   Namdeo S, Baghel RPS, Nayak S, *et al.* Effect of dietary supplementation of ginger, garlic and turmeric on humoral immune response, antioxidant property and carcass traits of broilers. 2022.

[51]   El-Deek AA, Attia YA, Maysa M, Hannfy M. Effect of anise (*Pimpinella anisum*), ginger(*Zingiber officinale* Roscoe) and fennel (*Foeniculum vulgare*) and their mixture on performance of broilers.

Arch Geflugelkd 2002; 67: 92-6.

[52]   Sies H. Oxidative stress: From basic research to clinical application. Am J Med 1991; 91(3): S31-8.
[http://dx.doi.org/10.1016/0002-9343(91)90281-2] [PMID: 1928209]

[53]   Addeo NF, Vozzo S, Secci G, *et al.* Different combinations of butchery and vegetable wastes on growth performance, chemical-nutritional characteristics and oxidative status of black soldier fly growing larvae. Animals (Basel) 2021; 11(12): 3515.
[http://dx.doi.org/10.3390/ani11123515] [PMID: 34944290]

[54]   Toghyani M, Mosavi S, Modaresi M, Landy N. Evaluation of kefir as a potential probiotic on growth performance, serum biochemistry and immune responses in broiler chicks. Anim Nutr 2015; 1(4): 305-9.
[http://dx.doi.org/10.1016/j.aninu.2015.11.010] [PMID: 29767062]

[55]   Shah M, Chand N, Khan RU, *et al.* Mitigating heat stress in broiler chickens using dietary onion (*Allium cepa*) and ginger (*Zingiber officinale*) supplementation. S Afr J Anim Sci 2023; 52(6): 811-8.
[http://dx.doi.org/10.4314/sajas.v52i6.07]

[56]   Sahoo N, Mishra SK, Swain RK, *et al.* Effect of turmeric and ginger supplementation on immunity, antioxidant, liver enzyme activity, gut bacterial load and histopathology of broilers. Indian J Anim Sci 2019; 89(7): 774-9.
[http://dx.doi.org/10.56093/ijans.v89i7.92046]

[57]   Huang SW, Frankel EN. Antioxidant activity of tea catechins in different lipid systems. J Agric Food Chem 1997; 45(8): 3033-8.
[http://dx.doi.org/10.1021/jf9609744]

[58]   Singh G, Marimuthu P, Heluani CS, Catalan C. DE-Heluani CS, Catalan C. Antimicrobial and antioxidant potentials of essential oil and acetone extract of *Myristica fragrans* Houtt. (aril part). J Food Sci 2005; 70(2): M141-8.
[http://dx.doi.org/10.1111/j.1365-2621.2005.tb07105.x]

[59]   Batiha GES, Beshbishy AM, Ikram M, *et al.* The pharmacological activity, biochemical properties, and pharmacokinetics of the major natural polyphenolic flavonoid: Quercetin. Foods 2020; 9(3): 374.
[http://dx.doi.org/10.3390/foods9030374] [PMID: 32210182]

[60]   Swelum AA, Hashem NM, Abdelnour SA, *et al.* Effects of phytogenic feed additives on the reproductive performance of animals. Saudi J Biol Sci 2021; 28(10): 5816-22.
[http://dx.doi.org/10.1016/j.sjbs.2021.06.045] [PMID: 34588896]

[61]   Abd El-Hack ME, El-Saadony MT, Salem HM, *et al.* Alternatives to antibiotics for organic poultry production: types, modes of action and impacts on bird's health and production. Poult Sci 2022; 101(4): 101696.
[http://dx.doi.org/10.1016/j.psj.2022.101696] [PMID: 35150942]

[62]   Abo El-Maaty H, Sherif S, Taha AE, *et al.* Effects of housing systems and feed additive on growth, carcass traits, liver function, oxidative status, thyroid function, and immune parameters of broilers. Poult Sci 2023; 102(12): 103121.
[http://dx.doi.org/10.1016/j.psj.2023.103121] [PMID: 37852054]

[63]   Aeschbach R, Löliger J, Scott BC, *et al.* Antioxidant actions of thymol, carvacrol, 6-gingerol, zingerone and hydroxytyrosol. Food Chem Toxicol 1994; 32(1): 31-6.
[http://dx.doi.org/10.1016/0278-6915(84)90033-4] [PMID: 7510659]

[64]   Kuo JM, Yeh DB, Pan BS. Rapid photometric assay evaluating antioxidative activity in edible plant material. J Agric Food Chem 1999; 47(8): 3206-9.
[http://dx.doi.org/10.1021/jf981351o] [PMID: 10552632]

[65]   Saeid JM, Mohamed AB, AL-Baddy MA. Effect of Aqueous Extract of Ginger (*Zingiber officinale*) on Blood Biochemistry Parameters of Broiler. Int J Poult Sci 2010; 9(10): 944-7.
[http://dx.doi.org/10.3923/ijps.2010.944.947]

[66] Tanabe M, Chen YD, Saito K, Kano Y. Cholesterol biosynthesis inhibitory component from *Zingiber officinale* Roscoe. Chemical and Pharmaceutical Bulletin, (Tokyo) 1993; 41: 710-713.

[67] Srinivasan K, Sambaiah K. The effect of spices on cholesterol 7 alpha-hydroxylase activity and on serum and hepatic cholesterol levels in the rat. Int J Vitam Nutr Res 1991; 61(4): 364-9.
[PMID: 1806542]

[68] Dou Y, Gregersen S, Zhao J, Zhuang F, Gregersen H. Morphometric and biomechanical intestinal remodeling induced by fasting in rats. Dig Dis Sci 2002; 47(5): 1158-68.
[http://dx.doi.org/10.1023/A:1015019030514] [PMID: 12018916]

[69] Yamauchi K, Buwjoom T, Koge K, Ebashi T. Histological intestinal recovery in chickens refed dietary sugar cane extract. Poult Sci 2006; 85(4): 645-51.
[http://dx.doi.org/10.1093/ps/85.4.645] [PMID: 16615348]

[70] Sittiya J, Yamauchi K. Growth performance and histological intestinal alterations of Sanuki Cochin chickens fed diets diluted with untreated whole-grain paddy rice. J Poult Sci 2014; 51(1): 52-7.
[http://dx.doi.org/10.2141/jpsa.0130042]

[71] Caspary WF. Physiology and pathophysiology of intestinal absorption. Am J Clin Nutr 1992; 55(1) (Suppl.): 299S-308S.
[http://dx.doi.org/10.1093/ajcn/55.1.299s] [PMID: 1728844]

[72] Ravindran V, Abdollahi MR. Nutrition and digestive physiology of the broiler chick: State of the art and outlook. Animals (Basel) 2021; 11(10): 2795.
[http://dx.doi.org/10.3390/ani11102795] [PMID: 34679817]

[73] Langhout DJ, Schutte JB, Van Leeuwen P, Wiebenga J, Tamminga S. Effect of dietary high-and low-methylated citrus pectin on the activity of the ileal microflora and morphology of the small intestinal wall of broiler chicks. Br Poult Sci 1999; 40(3): 340-7.
[http://dx.doi.org/10.1080/00071669987421] [PMID: 10475630]

[74] Yasar S, Forbes JM. Performance and gastro-intestinal response of broiler chickens fed on cereal grain-based foods soaked in water. Br Poult Sci 1999; 40(1): 65-76.
[http://dx.doi.org/10.1080/00071669987854] [PMID: 10405038]

[75] Modina SC, Aidos L, Rossi R, Pocar P, Corino C, Di Giancamillo A. Stages of gut development as a useful tool to prevent gut alterations in piglets. Animals (Basel) 2021; 11(5): 1412.
[http://dx.doi.org/10.3390/ani11051412] [PMID: 34069190]

[76] Moniello G, Ariano A, Panettieri V, *et al.* Intestinal morphometry, enzymatic and microbial activity in laying hens fed different levels of a *Hermetia illucens* larvae meal and toxic elements content of the insect meal and diets. Animals (Basel) 2019; 9(3): 86.
[http://dx.doi.org/10.3390/ani9030086] [PMID: 30857338]

[77] Garland CD. LEE A, Dickson MR. Segmented filamentous bacteria in the rodent small intestine: their colonization of growing animals and possible role in host resistance to *Salmonella*. Microb Ecol 1982; 8: 181-90.
[http://dx.doi.org/10.1007/BF02010451] [PMID: 24225812]

[78] Heczko U, Abe A, Finlay BB. *In vivo* interactions of rabbit enteropathogenic O103 with its host: an electron microscopic and histopathologic study. Microbes Infect 2000; 2(1): 5-16.
[http://dx.doi.org/10.1016/S1286-4579(00)00291-4] [PMID: 10717535]

[79] Sudarshan S, Fairoze N, Ruban SW, Badhe SR, Raghunath BV. Effect of aqueous extract and essential oils of ginger and garlic as immunostimulant in chicken meat. Research Journal of Poultry Sciences 2010; 3(3): 58-61.
[http://dx.doi.org/10.3923/rjpscience.2010.58.61]

[80] Ganiswarna SG. Farmakologi dan Terapi (Pharmacology and Teraphy). Medical Faculty, Indonesia University, Gaya Baru Jakarta, 1995; 471.

CHAPTER 10

Bee Pollen

Youssef A. Attia[1,2]**, Mohamed E. Abd El-Hack**[3,*]**, Mahmoud Alagawany**[3]**, Salem R. Alyileili**[4]**, Khalid A. Asiry**[1]**, Saber S. Hassan**[2]**, Asmaa Sh. Elnaggar**[2]**, Hany I. Habiba**[2]**, Shatha I. Alqurashi**[5]**, Asmaa F. Khafaga**[6] **and Maria Cristina de Oliveira**[7]

[1] *Sustainable Agriculture Production Research Group, Agriculture Department, Faculty of Environmental Sciences, King Abdulaziz University, Jeddah-21589, Saudi Arabia*

[2] *Animal and Poultry Production Department, Faculty of Agriculture, Damanhour University, Damanhour-22713, Egypt*

[3] *Poultry Department, Faculty of Agriculture, Zagazig University, Zagazig-44519, Egypt*

[4] *Department of Laboratory Analyses, College of food and Agriculture Sciences, United Arab Emirates University, Al Ain United Arab Emirates*

[5] *Department of Biological Science, College of Science, University of Jeddah, Jeddah-21589, Saudi Arabia*

[6] *Department of Pathology, Faculty of Veterinary Medicine, Alexandria University, Apis, Alexandria, 21944, Egypt*

[7] *Faculty of Veterinary Medicine, University of Rio Verde, Rio Verde, GO, Brazil*

Abstract: Bee Pollen (BP) is a mixture of nectar, salivary secretions from bees, and pollen grains collected from the flowers. It contains a wide range of nutrients, including proteins (10-40%), carbohydrates (13-55%), lipids (1-20%), vitamins (0.02-0.1%), minerals (0.5-3%), flavonoids (0.04-3%), and other bioactive substances such as phenolic compounds. BP has been reported to possess various therapeutic properties including antioxidant, anti-inflammatory, immunomodulatory, and antimicrobial activities. The chemical composition and bioactive substances in BP may differ significantly owing to factors such as plant species, nutritional status, environmental conditions, age, and vegetation during the flowering period. BP has been shown to have beneficial effects on human health, including the prevention of prostate problems, arteriosclerosis, and tumors. In animal science, BP supplementation has been evaluated primarily in poultry with encouraging results. BP can improve the cell immune response, antibody production speed and reinforce the immunological system. The positive effects of BP on animal productive performance may be due to its nutritive value, appetite-stimulant properties, and the presence of digestive enzymes. In domestic animals such as sheep, broilers, rabbits, and quails, supplementation with BP

* **Corresponding author Mohamed E. Abd El-Hack:** Poultry Department, Faculty of Agriculture, Zagazig University, Zagazig-44519, Egypt; E-mail: dr.mohamed.e.abdalhaq@gmail.com

has been reported to improve the immune response, increase feed digestibility, reduce oxidative stress, and improve animal performance. This chapter emphasizes the use of Bee Pollen in livestock nutrition as a feed supplement to improve productive performance as an eco-friendly alternative to antibiotics.

Keywords: Antioxidants, Animal performance, Bioactive components, Bee Pollen, Growth promoters, Livestock.

INTRODUCTION

Ordinary and organic Bee Pollen (BP) are shown in Figs. (**1** and **2**). BP (Fig. **1**) is one of the purest and potential natural foods ever discovered and has shown incredible nutritional and medicinal value [1 - 3]. Pollen collected by bees is superior to that collected directly from flowering plants, and bees are highly discriminated about selecting the best pollen from the millions of grains present, and those only rich in all the nutrients, particularly nitrogenous substances [4 - 7]. Bees mix the pollen particles with sticky material secreted from their stomachs, allowing the BP to adhere to their rear legs in "pollen baskets" to transport it to their hives safely [8, 9]. The color and shape of BP, as indicated in Fig. (**1**) reveal the plant species from which it was obtained and the particular geographical area. Although pollen color is normally neglected, it ranges from golden yellow to black, according to the source [3, 10 - 12].

Fig. (1). Bee Pollen. Source: Contribution of conger design by Pixabay [13].

Fig. (2). Organic Bee Pollen. Source: Contribution of conger design by Pixabay [13].

BEE POLLEN

Pollen contains different coloring agents and pigments, only a small amount of which has been isolated [3, 14]. Certain pigments are fat soluble, whereas others are water-soluble [7]. The chemical compositions and bioactive components of BP are listed in Table **1**.

Table 1. Nutritive composition of Bee Pollen.

Component	Amount	Component	Rate
Energy (kcal/kg)	2.46	Ni (ppm)	4.50
Protein (%)	23.7	B_1 (ppm)	9.40
Carbohydrate (%).	27	Niacin (ppm)	157
Lipid (%).	4.8	B_2 (ppm)	18.6
P (%)	0.53	B_6 (ppm)	9
K (%)	0.58	Pantothenate (ppm)	28
Na (%)	0.044	Folic acid (ppm)	5.20
Ca (%)	0.225	Biotin (ppm)	0.32
Mg (%)	0.148	Vit. C (ppm)	350
Zn (ppm)	87	Carotenes (ppm)	95

(Table 1) cont.....

Component	Amount	Component	Rate
Cu (ppm)	14	Amino Acids (%)	10-35
Fe (ppm)	140		

Source: Schmidt [15].

Bee Pollen is a mixture of nectar, salivary secretions from bees, and pollen grains collected from flowers [3, 4], and presents 3-8% water, 10-40% proteins, 13-55% carbohydrates, 1-20% lipids, 0.02-0.1% vitamins, 0.5-3% minerals, 0.04-3% flavonoids and other constituents like antibiotic substances and resins [16, 17]. In other studies, BP was found to have 5-60% proteins,13-55% sugars, 0.3-20% crude fiber, 4-7% lipids, and 3-8% minerals and phenolic compounds, mainly flavonoids [6]. The chemical composition of the BP is presented in Fig. (3).

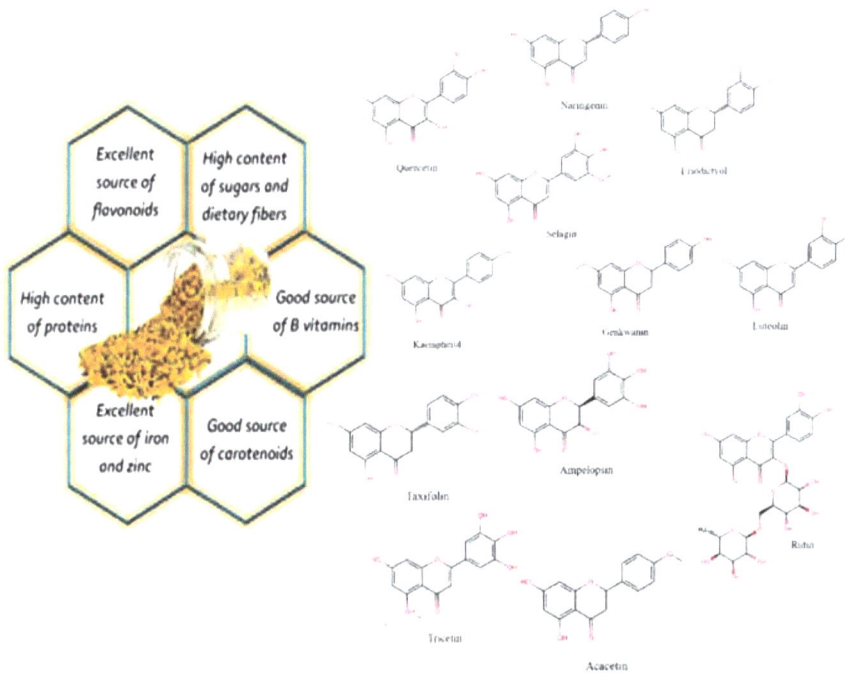

Fig. (3). Chemical structure and beneficial nutritional constituents of Bee Pollen.

Bee Pollen has been reported to possess various therapeutic properties, including antioxidant, anti-inflammatory, immunomodulatory, and antimicrobial activities. The chemical composition and bioactive substances in BP may differ significantly depending on the plant species from which the pollen was gathered [10, 17], nutritional status, environmental conditions, age, and vegetation during the flowering period [6, 10, 12].

Bee Pollen is beneficial as a healthy food with a wide variety of therapeutic and nutritional possessions [4, 18], and has beneficial effects on human health and the prevention of prostate problems [14, 19], arteriosclerosis [20], and tumors [21, 22]. BP has been reported to accelerate mitosis, promote tissue repair, and enhance the elimination of toxic substances [23]. In addition, it has antimicrobial, anti-inflammatory, antimutagenic [5, 24, 25], antifungal [24, 26], antioxidant [6, 21], anti-allergic [27], antiviral, hypolipidemic, hypoglycemic, and immunostimulatory properties [28].

Bee Pollen can also improve the cell immune response and antibody production rate, reinforcing the immunological system [10, 11, 29]. However, Attia *et al.* [30] reported that BP did not affect the mortality and total bacterial count of growing rabbits or the survivability of broiler chickens. To increase the profitability of livestock farming, it is essential to improve the productivity and survival of growing animals [12, 31]. Therefore, antibiotics have been used in diets for growing animals to decrease the incidence of diseases and act as growth propolis [7, 10, 32].

The indiscriminate use of antibiotics contributes to an increase in the appearance of bacteria resistant to their effects, representing a significant threat to the health of animals and humans [33, 34]. Consequently, the use of antibiotics as growth enhancers has been prohibited in several countries, representing a major challenge for animal meat and egg producers [12, 30]. For these reasons, interest in the search for new natural alternatives to antibiotics and synthetic antioxidants has increased in recent years [6, 12]. The health benefits of BP are shown in Fig. (**4**).

Among the currently available natural alternatives are products derived from bees (*Apis mellifera*), such as BP. These products contain various bioactive metabolites with pharmaceutical properties [3, 35, 36]. In animal science, the effects of BP supplementation have been evaluated primarily in poultry with encouraging results [5, 34, 36].

Effects of Bee Pollen

Productive Performance

The positive effects of BP on animal productive performance could be due to its nutritive value, which is rich in nutrients such as amino acids, fats, vitamin C, oligoelements, enzymes, and flavonoids, which are antioxidant factors that are important for cell and intestinal mucosa differentiation [37, 3], responsible for nutrient absorption. In addition, BP is considered an appetite-stimulant, as demonstrated by Peric *et al.* [40] and Nemauluma *et al.* [41] in broilers.

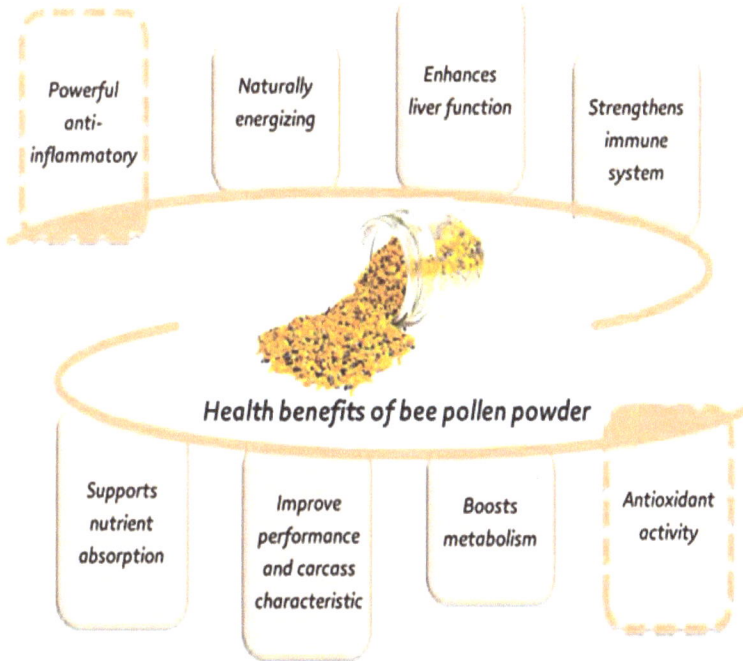

Fig. (4). Health benefits of bee-pollen.

Furthermore, the presence of digestive enzymes in BP [42] and an increase in the number of lactobacilli that acidify the intestine and increase mineral absorption and amylase, protease, and phytate secretion [43] collaborated to improve nutrient use by birds. In domestic animals such as sheep, broilers, rabbits, and quails, it has been reported that supplementation with BP improves the immune response [2], increases feed digestibility [44], reduces oxidative stress [45], and improves animal performance [11, 36].

Improvements in poultry production performance have been reported by several researchers. For example, Wang *et al.* [46] noted that adding 1.5% dietary BP increased the body weight of broilers by 35.1% compared with the control group. Al-Kahtani *et al.* [2] showed a positive effect of dietary BP (1 g/kg) on the growth and feed intake of broilers at 42 days of age compared to non-supplemented birds. A similar effect was observed by Hosseini *et al.* [47], Prakatur *et al.* [48], and Petricevic *et al.* [49]. In a study of Japanese quails, Oliveira *et al.* [50] reported an increase in feed intake, laying rate, and egg mass due to the inclusion of 1% and 1.5% dietary BP.

Rabbits can also benefit from the inclusion of BP in their diets. Attia *et al.* [14] showed that 200 mg/kg of body weight of BP resulted in better weight gain and lower feed intake in rabbits during pregnancy compared to the dose of the control group. Abel-Hamid and El-Tarabany [51] and Omar *et al.* [52] improved the final body weight, weight gain, feed intake, and feed conversion rate of rabbits due to the administration of 350 and 500 mg/kg body weight, respectively.

Positive effects of BP on fish diets have been previously reported. Abbas *et al.* [53] reported an improvement in specific growth rate, body weight, and feed conversion rate in females but not in males of Nile tilapia-fed diets with 2.5% BP. Nowosad *et al.* [54] fed African catfish with diets containing 1%, 2%, and 3% of BP. The fish presented higher final weight, growth rate, and specific growth rate than the control group.

However, Oliveira *et al.* [55] did not observe an influence of 1.5% dietary BP on the productive performance of broilers. Similarly, Attia *et al.* [10] and Sierra-Galicia *et al.* [7] reported no effect of 300 mg in the diet and 500 mg/kg body weight, respectively, on the productive performance of growing rabbits.

Carcass Characteristics

Bee Pollen is rich in amino acids, such as leucine, which may impact anabolic functions in muscle cells [56]. In addition, antioxidant compounds present in BP may increase the ability of leucine to stimulate muscle protein synthesis [57]. Salles *et al.* [58] showed that BP restored the weight of tissues and organs in feed-restricted old rats when they were refeeding, with a reduction in the subcutaneous adipose tissues.

Studies have been conducted to evaluate the effects of BP on carcass characteristics, particularly in poultry. Hosseini *et al.* [47] pointed out that 20 g/kg BP was enough to increase the spleen and bursa weight in broilers reared under heat stress. Abood *et al.* [59] and Nemauluma *et al.* [41] noted increased carcass dressing and weights of the proventriculus, liver, gizzard, and small intestine.

Rabbit research by Dias *et al.* [38] showed that the use of 1 g/d was not able to improve the carcass characteristics of growing rabbits, except for an increase in the weight of the gut. However, Omar *et al.* [52] reported that rabbits reared under high stocking density and receiving 500 mg/kg body weight of BP had higher slaughter weight, carcass yield, liver and kidney weights, and a reduction in perirenal fat. The same effect was verified by Zeedan *et al.* [60] using 200, 500, and 700 mg/kg body weight BP and Sierra-Galicia *et al.* [12] using 500 mg/kg BP in rabbit diets.

There are contradictory results in the literature, such as those by Abdel-Hamid and El-Tarabany [51], who reported that 250 and 350 mg of dietary BP did not affect the weight and yield of rabbit carcasses compared to the values obtained with non-supplemented rabbits. Attia *et al.* [10, 32] also revealed that the use of 150 to 300 mg BP did not affect the carcass traits or the weight of the liver, spleen, heart, and lungs of rabbits. The lack of changes in the carcass of Japanese quails was seen by Canogullar *et al.* [61] using BP at 0, 5, 10, and 20 g/kg diet and by Sevin [62] using BP at 2.5, 5, and 10 g/kg. Petrićević *et al.* [49] also reported the absence of an effect of 0.25, 0.5, 0.75, and 1% BP on the carcass traits of broilers.

Farag and El-Rayes [63] found that the carcass percentage increased significantly (p<0.01) in groups fed bee-pollen diets compared to that in the control group. The highest values (%) of the liver (2.07) and gizzard (2.21) were obtained from broilers fed the control diet, while the highest values (%) of the percentage of the heart (0.47), thymus (0.40), bursa (0.11), and spleen (0.15) were recorded in chicks fed Bee Pollen at 0.6% level. Recently, Sierra-Galicia *et al.* (2023) found that Bee Pollen increased the carcass yield of rabbits (< 0.001) after reviewing recent publications on rabbits.

Blood Profile

Bee Pollen may positively influence the blood profiles of animals. For example, BP exerts anti-hyperglycemic effects by modulating glucose uptake, inhibiting the activities of the enzymes α-amylase and α-glucosidase [64], stimulating glucokinase, an enzyme that promotes hepatic glucose uptake and glycogen storage [65], stimulating the production and function of pancreatic β-cells [66], and reducing blood glucose levels, as shown by Rahayu *et al.* [67] and Setyawan *et al.* [68] in alloxan-induced diabetic rats. However, this effect is not commonly observed in farm animals [12, 14, 32, 62, 69].

In addition, BP may have a hypocholesterolemic effect because naringin, one of its compounds, may block the activity of HMGCoA reductase, an enzyme that participates in the metabolic pathway to produce cholesterol [70], reducing cholesterol in the serum, as described by Attia *et al.* [1, 14] and by Abdel-Hamid and El-Tarabany [51]. in rabbits fed diets containing BP. However, Sevin [62] did not verify the positive effect of BP on the blood levels of cholesterol and triglycerides in Japanese quails.

In addition, its antioxidant properties help increase the antioxidant defense in the body and reduce oxidative stress markers [71]. Dietary BP at 1 g/kg increases the total antioxidant capacity and the levels of superoxide dismutase and catalase in broilers [2].

Adequate levels of urea and creatinine can be used as biomarkers of renal function, and serum levels of hepatic enzymes are generally used as markers of liver disorders [12]. Owing to its antioxidant effect, BP can maintain membrane cell integrity in the kidneys and liver, improving renal and hepatic function.

Attia *et al*. [1] showed that NZW rabbits receiving 200 mg/kg body weight of Bee Pollen had lower blood levels of creatinine, urea, AST, and ALT, compared to the non-supplemented animals, a control group. In this study, BP increased the plasma glucose, total protein, albumin, and globulin levels. The same effect was observed in broilers by Attia *et al*. [30] and rabbits [14], who demonstrated that dietary BP decreased the levels of triglycerides, total cholesterol, urea, creatinine, and aspartate aminotransferase.

The antioxidant activity of BP protects the cell membrane from free radical attack. This is important for maintaining the blood cell count within the reference range for each species. The use of BP in animal diets has been associated with an increase in the red blood cell count and hemoglobin concentration in rabbits [30].

Wang *et al*. [72] showed that broilers (1-42 d of age) supplemented with BP at 1.5% diet had greater relative weight of thymus, bursa, and spleen than the control group, indicating an improvement of the immune response of the birds. Attia *et al*. [32] found that Bee Pollen did not significantly influence most of the haematological parameters of growing rabbits except for an increase in phagocyte activity. In addition, the use of BP in rabbit diets may increase red blood cells and hemoglobin content [14, 30]. Bee Pollen supplied to patients with anemia may mitigate the negative effects of iron deficiency in rats [20].

More recently, Attia *et al*. [4] found that BP supplied continuously or intermittently improved immunity and antioxidant enzyme levels in broilers. The use of BP in a chick's diet (300 mg/kg), in a continuous or intermittent manner, increases white blood cell counts as well as the percentage of lymphocytes and heterophils, improving the immunity of the bird.

Al-Khatani *et al*. [2] reported that dietary BP (1 g/kg) increased leukocyte cell viability and the levels of T and B lymphocytes, IgA, and IgM in broilers at 42 days of age compared to non-supplemented birds. Hassan *et al*. [3] obtained similar results in rabbits with improvement in white and red blood cell counts and the levels of lymphocytes, hemoglobin, and packed cell volume due to the use of 400 mg /kg body weight.

CONCLUSION AND REMARKS

Bee Pollen is a naturally occurring product that has recently been used for animal nutrition and treatment. It was concluded from the previous studies that adding BP to the food at a rate of 1 to 20 g/kg enhances the reproductive, immunological, and productive status of animals. Thus, using BP as a feed supplement could provide a proper nutritional strategy for animal production. Additionally, BP has many priceless medicinal characteristics since ancient times. Scientists have proven the biological and therapeutic activities of some of these magnificent superfoods. However, they have just started to learn about their various health benefits, and there are only a few reliable sources of support for those claims. Research and development in this area typically focus on complete BP materials, rather than employing individual components. However, the complexity and compositional diversity of these products makes standardization necessary for safe and predictable animal applications. However, little is known about BP's chemistry and bioactive components of BPs.

Additionally, specific substances from honeybee products have not yet been used for medical treatment. Additional studies (*in vitro*, in animals, and in clinical trials) and validation are required to demonstrate any practical effects and the mechanism of action of natural BP as well as isolated molecules. Studies aimed at improving our knowledge of their mechanisms of action continue to be crucial for creating applications of BP products for animal use. In conclusion, Bee Pollen (BP) has emerged as a promising natural supplement in animal nutrition, offering a wide range of benefits for livestock health and productivity. The rich nutritional profile of BP, which includes proteins, carbohydrates, lipids, vitamins, minerals, and bioactive compounds, contributes to its diverse therapeutic properties. These include antioxidant, anti-inflammatory, immunomodulatory, and antimicrobial effects that have been demonstrated in various animal studies.

The application of BP to livestock nutrition has shown encouraging results, particularly in poultry, rabbits, and fish. Improvements have been observed in production performance, carcass characteristics, blood profiles, and immune responses. However, it is important to note that the results can vary depending on factors such as dosage, animal species, and environmental conditions, emphasizing the need for further research to fully understand its mechanisms of action and standardize its use across different livestock species. More studies are required to explore the specific bioactive components of BP and their individual effects on animal health and performance.

As the livestock industry continues to seek sustainable and natural alternatives to synthetic additives, Bee Pollen is a promising candidate. Its multifaceted benefits and natural origin make it an attractive option for improving animal welfare and productivity, while addressing concerns about antibiotic resistance and consumer demand for more natural food production methods.

IMPLICATIONS

- BP offers a potential eco-friendly substitute for antibiotics as a growth promoter in animal feed, addressing concerns about antibiotic resistance and meeting consumer demands for more natural food production methods.
- Fortification with BP has shown promising results in enhancing productive performance, including increasing body weight, feed intake, and feed conversion rates in various livestock species, particularly poultry and rabbits.
- BP's immunomodulatory properties of BP can strengthen the immune system of animals, potentially leading to improved disease resistance and overall health.
- Antioxidant compounds in BP may help reduce oxidative stress in animals, contributing to better overall health and potentially improving meat quality.
- BP's rich nutritional profile of BP, including proteins, vitamins, and minerals, can supplement animal diets and potentially improve nutrient utilization.
- Some studies have shown positive effects on carcass characteristics, which can lead to improved meat quality and yield.
- The variability in BP composition based on plant sources suggests the possibility of developing specialized BP supplements for specific livestock needs.
- BP fortification consistently improves animal performance and health and can lead to increased profitability for livestock producers.
- The differences in results across studies highlight the necessity for more comprehensive research to standardize BP use in animal nutrition and to fully understand its mechanisms of action.
- The application of BP as a natural feed additive aligns with sustainable agricultural practices and could contribute to environmentally friendly livestock production.
- The implications of BP in animal nutrition showed the potential to enhance livestock nutrition and management practices, although more research is needed to optimize its application across different animal species and production systems, such as traditional and organic systems.

REFERENCES

[1]   Attia YA, Al-Hanoun A, Bovera F. Effect of different levels of Bee Pollen on performance and blood profile of New Zealand White bucks and growth performance of their offspring during summer and winter months. J Anim Physiol Anim Nutr (Berl) 2011; 95(1): 17-26.
[http://dx.doi.org/10.1111/j.1439-0396.2009.00967.x] [PMID: 20455966]

[2]   AL-Kahtani SN, Alaqil AA, Abbas AO. Modulation of antioxidant defense, immune response, and growth performance by inclusion of propolis and Bee Pollen into broiler diets. Animals (Basel) 2022;

12(13): 1658.
[http://dx.doi.org/10.3390/ani12131658] [PMID: 35804557]

[3]   Hassan SS, Shahba HA, Mansour M. Influence of using date palm pollen or Bee Pollen on some blood biochemical metabolites, semen characteristics and subsequent reproductive performance of v-line male rabbits. Egyptian Journal of Rabbit Science 2022; 32(1): 19-39.
[http://dx.doi.org/10.21608/ejrs.2022.232781]

[4]   Attia YA, Al-Khalaifah H, Ibrahim MS, Al-Hamid AEA, Al-Harthi MA, Elnaggar Sh. El-Naggar ShA. Blood haematological and biochemical constituents, antioxidant enzymes, immunity and lymphoid organs of broiler chicks supplemented with propolis, Bee Pollen and mannan oligosaccharides continuously or intermittently. Poult Sci 2017; 96(12): 4182-92.
[http://dx.doi.org/10.3382/ps/pex173] [PMID: 29053876]

[5]   Lika E, Kostić M, Vještica S, Milojević I, Puvača N. Honeybee and plant products as natural antimicrobials in enhancement of poultry health and production. Sustainability (Basel) 2021; 13(15): 8467.
[http://dx.doi.org/10.3390/su13158467]

[6]   Martinello M, Mutinelli F. Antioxidant activity in bee products: a review. Antioxidants 2021; 10(1): 71.
[http://dx.doi.org/10.3390/antiox10010071] [PMID: 33430511]

[7]   Sierra-Galicia MI, Rodríguez-de Lara R, Orzuna-Orzuna JF, *et al.* Supplying Bee Pollen and propolis to growing rabbits: effects on growth performance, blood metabolites, and meat quality. Life (Basel) 2022; 12(12): 1987.
[http://dx.doi.org/10.3390/life12121987] [PMID: 36556352]

[8]   Hoffman GD, Lande C, Rao S. A novel pollen transfer mechanism by honey bee foragers on highbush blueberry (Ericales: Ericaceae). Environ Entomol 2018; 47(6): 1465-70.
[http://dx.doi.org/10.1093/ee/nvy162] [PMID: 30452583]

[9]   Olszewski K, Dziechciarz P, Trytek M, Borsuk G. A scientific note on the strategy of wax collection as rare behavior of *Apis mellifera.* Apidologie (Celle) 2022; 53(4): 40.
[http://dx.doi.org/10.1007/s13592-022-00948-z]

[10]   Attia YA, Bovera F, Abd-Elhamid AEHE, *et al.* Evaluation of the carryover effect of antibiotic, Bee Pollen and propolis on growth performance, carcass traits and splenic and hepatic histology of growing rabbits. J Anim Physiol Anim Nutr (Berl) 2019; 103(3): 947-58.
[http://dx.doi.org/10.1111/jpn.13068] [PMID: 30714248]

[11]   Attia YA, Bovera F, Abd El-Hamid AE, *et al.* Bee pollen and propolis as dietary supplements for rabbit: Effect on reproductive performance of does and on immunological response of does and their offspring. Journal of Animal Physiology and Animal Nutrition 2019aab; 103: 959-968.
[http://dx.doi.org/10.1111/jpn.13069]

[12]   Sierra-Galicia MI, Rodríguez-de Lara R, Orzuna-Orzuna JF, Lara-Bueno A, Ramírez-Valverde R, Fallas-López M. Effects of supplementation with Bee Pollen and propolis on growth performance and serum metabolites of rabbits: a meta-analysis. Animals (Basel) 2023; 13(3): 439.
[http://dx.doi.org/10.3390/ani13030439] [PMID: 36766327]

[13]   Pixabay. Bee pollen. Contribution of congerdesign by Pixabay. 2017. Available from: https://pixabay.com/pt/photos/p%C3%B3len-de-abelha-p%C3%B3len-2549125

[14]   Attia YA, Bovera F, El-Tahawy WS, El-Hanoun AM, Al-Harthi MA, Habiba HI. Productive and reproductive performance of rabbits does as affected by Bee Pollen and/or propolis, inulin and/or mannan-oligosaccharides. World Rabbit Sci 2015; 23(4): 273-82.
[http://dx.doi.org/10.4995/wrs.2015.3644]

[15]   Schmidt JO. Bee product chemical - composition and application. In: Mizrahi A, Lensky Y, Eds. Bee products - properties, applications, and apitherapy. New York: Springer Nautre 1997; pp. 15-26.
[http://dx.doi.org/10.1007/978-1-4757-9371-0_2]

[16]   Carpes ST, Begnini R, Alencar SM, Masson ML. Study of preparations of Bee Pollen extracts, antioxidant and antibacterial activity. Cienc Agrotec 2007; 31(6): 1818-25.
[http://dx.doi.org/10.1590/S1413-70542007000600032]

[17]   Taha EKA. Chemical composition and amounts of mineral elements in honeybee-collected pollen in relation to botanical origin. J Apic Sci 2015; 59: 75-81.
[http://dx.doi.org/10.1515/jas-2015-0008]

[18]   Yamaguchi M, Hamamoto R, Uchiyama S, Ishiyama K, Hashimoto K. Anabolic effects of Bee Pollen *Cistus ladaniferus* extract on bone components in the femoral-diaphyseal and -metaphyseal tissues of rats *in vitro* and *in vivo*. J Health Sci 2006; 52(1): 43-9.
[http://dx.doi.org/10.1248/jhs.52.43]

[19]   Shoskes DA. Phytotherapy in chronic prostatitis. Urology 2002; 60(6) (Suppl.): 35-7.
[http://dx.doi.org/10.1016/S0090-4295(02)02383-X] [PMID: 12521591]

[20]   Rzepecka-Stojko A, Stojko J, Jasik K, Buszman E. Anti-atherogenic activity of polyphenol-rich extract from Bee Pollen. Nutrients 2017; 9(12): 1369.
[http://dx.doi.org/10.3390/nu9121369] [PMID: 29258230]

[21]   Wan Omar WA, Azhar NA, Harif Fadzilah N, Nik Mohamed Kamal NNS. Bee Pollen extract of Malaysian stingless bee enhances the effect of cisplatin on breast cancer cell lines. Asian Pac J Trop Biomed 2016; 6(3): 265-9.
[http://dx.doi.org/10.1016/j.apjtb.2015.12.011]

[22]   Arung ET, Ramadhan R, Khairunnisa B, *et al.* Cytotoxicity effect of honey, Bee Pollen, and propolis from seven stingless bees in some cancer cell lines. Saudi J Biol Sci 2021; 28(12): 7182-9.
[http://dx.doi.org/10.1016/j.sjbs.2021.08.017] [PMID: 34867021]

[23]   Morais M, Moreira L, Feás X, Estevinho LM. Honeybee-collected pollen from five Portuguese Natural Parks: Palynological origin, phenolic content, antioxidant properties and antimicrobial activity. Food Chem Toxicol 2011; 49(5): 1096-101.
[http://dx.doi.org/10.1016/j.fct.2011.01.020] [PMID: 21291944]

[24]   Kacániová M, Vukovic N, Chlebo R, *et al.* The antimicrobial activity of honey, Bee Pollen loads and beeswax from Slovakia. Arch Biol Sci 2012; 64(3): 927-34.
[http://dx.doi.org/10.2298/ABS1203927K]

[25]   Pascoal A, Rodrigues S, Teixeira A, Feás X, Estevinho LM. Biological activities of commercial Bee Pollens: Antimicrobial, antimutagenic, antioxidant and anti-inflammatory. Food Chem Toxicol 2014; 63: 233-9.
[http://dx.doi.org/10.1016/j.fct.2013.11.010] [PMID: 24262487]

[26]   Garcia M, Peres-Aquillue C, Juan T, Juan MI, Herrera A. Pollen analysis and antibacterial activity of spanish honeys. Food Sci Technol Int 2001; 7: 155-8.
[http://dx.doi.org/10.1177/108201320100700208]

[27]   Moita E, Sousa C, Andrade P, *et al.* Effects of *Echium plantagineum* L. Bee Pollen on basophil degranulation: relationship with metabolic profile. Molecules 2014; 19(7): 10635-49.
[http://dx.doi.org/10.3390/molecules190710635] [PMID: 25054443]

[28]   Komosinska-Vassev K, Olczyk P, Kaźmierczak J, Mencner L, Olczyk K. Bee Pollen: chemical composition and therapeutic application. Evid Based Complement Alternat Med 2015; 2015: 1-6.
[http://dx.doi.org/10.1155/2015/297425] [PMID: 25861358]

[29]   Song YF, Wang J, Li SH, Shang CF. Effect of beepollen on the development of digestive gland of broilers. China Animal Husbandry and Veterinary Medicine 2005; 37: 14-7.

[30]   Attia YA, Al-Hamid AEA, Ibrahim MS, Al-Harthi MA, Bovera F, Elnaggar Sh. Productive performance, biochemical and hematological traits of broiler chickens supplemented with propolis, Bee Pollen, and mannan oligosaccharides continuously or intermittently. Livest Sci 2014; 164: 87-95.
[http://dx.doi.org/10.1016/j.livsci.2014.03.005]

[31]   Hashem NM, Abd El-Hady AM, Hassan OA. Inclusion of phytogenic feed additives comparable to vitamin E in diet of growing rabbits: Effects on metabolism and growth. Ann Agric Sci 2017; 62(2): 161-7.
[http://dx.doi.org/10.1016/j.aoas.2017.11.003]

[32]   Attia YA, El-Hanoun AM, Bovera F, Monastra G, El-Tahawy WS, Habiba HI. Growth performance, carcass quality, biochemical and haematological traits and immune response of growing rabbits as affected by different growth promoters. J Anim Physiol Anim Nutr (Berl) 2014; 98(1): 128-39.
[http://dx.doi.org/10.1111/jpn.12056] [PMID: 23419029]

[33]   Hashem NM, Hassanein EM, Simal-Gandara J. Improving reproductive performance and health of mammals using honeybee products. Antioxidants 2021; 10(3): 336.
[http://dx.doi.org/10.3390/antiox10030336] [PMID: 33668287]

[34]   Huang P, Cui X, Wang Z, *et al.* Effects of *Clostridium butyricum* and a bacteriophage cocktail on growth performance, serum biochemistry, digestive enzyme activities, intestinal morphology, immune responses, and the intestinal microbiota in rabbits. Antibiotics (Basel) 2021; 10(11): 1347.
[http://dx.doi.org/10.3390/antibiotics10111347] [PMID: 34827285]

[35]   Abdelnour SA, Abd El-Hack ME, Alagawany M, Farag MR, Elnesr SS. Beneficial impacts of Bee Pollen in animal production, reproduction and health. J Anim Physiol Anim Nutr (Berl) 2019; 103(2): 477-84.
[http://dx.doi.org/10.1111/jpn.13049] [PMID: 30593700]

[36]   Sadarman S, Erwan E, Irawan A, *et al.* Propolis supplementation affects performance, intestinal morphology, and bacterial population of broiler chickens. S Afr J Anim Sci 2021; 51(4): 477-87.
[http://dx.doi.org/10.4314/sajas.v51i4.8]

[37]   Wang J, Song Y, Li S, Wang Q. Effect of Bee Pollen on development of small intestine in broilers. Chin J Vet Sci Technol 2005; 35: 484-8.

[38]   Dias DMB, Oliveira MC, Silva DM, Bonifácio NP, Claro DDC, Marchesin WA. <b>Bee Pollen supplementation in diets for rabbit does and growing rabbits. Acta Sci Anim Sci 2013; 35(4): 425-30.
[http://dx.doi.org/10.4025/actascianimsci.v35i4.18950]

[39]   Aylanc V, Falcão SI, Vilas-Boas M. Bee Pollen and bee bread nutritional potential: Chemical composition and macronutrient digestibility under *in vitro* gastrointestinal system. Food Chem 2023; 413: 135597.
[http://dx.doi.org/10.1016/j.foodchem.2023.135597] [PMID: 36791664]

[40]   Peric L, Zikic D, Lukic M. Application of alternative growth promoters in broiler production. Biotechnol Anim Husb 2009; 25(5-6-1): 387-97.
[http://dx.doi.org/10.2298/BAH0906387P]

[41]   Nemauluma MFD, Manyelo TG, Ng'ambi JW, Malematja EM, Kolobe SD. Effects of bee pollen inclusion on the performance and gut morphology of Ross 308 broiler chickens. Brazilian Journal of Poultry Science 2023; 25: eRBCA-2022-1632.
[http://dx.doi.org/10.1590/1806-9061-2022-1632]

[42]   Wang J, Gu Y, Li S, Fang ZS, Feng BM. Effect of Bee Pollen on histological structure of digestive organ of layer. Journal of Anhui University Science and Tecnhology 2006; 20: 1-6.

[43]   Hamidreza T, Fatemeh T, Hossein M, Mojtaba Z, Mahmood S, Parvin S. Potential probiotic of *Lactobacillus johnsonii* LT171 for chicken nutrition. Afr J Biotechnol 2009; 8(21): 5833-7.
[http://dx.doi.org/10.5897/AJB09.1062]

[44]   Morsy AS, Soltan YA, El-Zaiat HM, Alencar SM, Abdalla AL. Bee propolis extract as a phytogenic feed additive to enhance diet digestibility, rumen microbial biosynthesis, mitigating methane formation and health status of late pregnant ewes. Anim Feed Sci Technol 2021; 273: 114834.
[http://dx.doi.org/10.1016/j.anifeedsci.2021.114834]

[45]   Cécere BGO, da Silva AS, Molosse VL, *et al.* Addition of propolis to milk improves lactating lamb's

growth: Effect on antimicrobial, antioxidant and immune responses in animals. Small Rumin Res 2021; 194: 106265.
[http://dx.doi.org/10.1016/j.smallrumres.2020.106265]

[46]   Wang J, Li S, Wang Q, Xin B, Wang H. Trophic effect of Bee Pollen on small intestine in broiler chickens. J Med Food 2007; 10(2): 276-80.
[http://dx.doi.org/10.1089/jmf.2006.215] [PMID: 17651063]

[47]   Hosseini S, Vakili Azghandi M, Ahani S, Nourmohammadi R. Effect of Bee Pollen and propolis (bee glue) on growth performance and biomarkers of heat stress in broiler chickens reared under high ambient temperature. J Anim Feed Sci 2016; 25(1): 45-51.
[http://dx.doi.org/10.22358/jafs/65586/2016]

[48]   Prakatur I, Domaćinović M, Lachner B, Steiner Z, Galović D, Miškulin I. Performance indicators of broilers fed propolis and Bee Pollen additive. Poljoprivreda (Osijek) 2019; 25(1): 69-75.
[http://dx.doi.org/10.18047/poljo.25.1.10]

[49]   Petricevic V, Lukic M, Skrbic Z, et al. Production parameters, microbiological composition of intestines and slaughter performance of broilers fed with Bee Pollen. Zuchtungskunde 2022; 94: 36-46.

[50]   Oliveira MC, Souza RG, Dias DMB, Gonçalves BN. Bee Pollen improves productivity of laying Japanese quails. Rev Bras Saúde Prod Anim 2020; 21: e212135020.
[http://dx.doi.org/10.1590/s1519-99402121352020]

[51]   Abdel-Hamid TM, El-Tarabany MS. Effect of Bee Pollen on growth performance, carcass traits, blood parameters, and the levels of metabolic hormones in New Zealand White and Rex rabbits. Trop Anim Health Prod 2019; 51(8): 2421-9.
[http://dx.doi.org/10.1007/s11250-019-01961-8] [PMID: 31187406]

[52]   Omar M, Hassan F, El-Shahat M. The effects of Bee Pollen on performance and economic efficiency of New Zealand White rabbits reared under high stocking density. Damanhour Journal of Veterinary Sciences 2020; 5(1): 18-23.
[http://dx.doi.org/10.21608/djvs.2020.156538]

[53]   Abbas AA, El-Asely AM, Kandiel MMM. Turk J Fish Aquat Sci 2012; 12(4): 851-9.
[http://dx.doi.org/10.4194/1303-2712-v12_4_13]

[54]   Nowosad J, Jasiński S, Arciuch-Rutkowska M, et al. Effects of Bee Pollen on growth performance, intestinal microbiota and histomorphometry in African catfish. Animals (Basel) 2022; 13(1): 132.
[http://dx.doi.org/10.3390/ani13010132] [PMID: 36611741]

[55]   Oliveira MC, Loch FC, Silva DM, Martins PC, Teixeira AS, Claro DC. Use of Bee Pollen in broiler diet. Rev Mex Cienc Pecu 2015; 6: 263-76. [https://www.scielo.org.mx/pdf/rmcp/v6n3/v6n3a2.pdf].

[56]   Le Plénier S, Walrand S, Noirt R, Cynober L, Moinard C. Effects of leucine and citrulline versus non-essential amino acids on muscle protein synthesis in fasted rat: a common activation pathway? Amino Acids 2012; 43(3): 1171-8.
[http://dx.doi.org/10.1007/s00726-011-1172-z] [PMID: 22160257]

[57]   Marzani B, Balage M, Vénien A, et al. Antioxidant supplementation restores defective leucine stimulation of protein synthesis in skeletal muscle from old rats. J Nutr 2008; 138(11): 2205-11.
[http://dx.doi.org/10.3945/jn.108.094029] [PMID: 18936220]

[58]   Salles J, Cardinault N, Patrac V, et al. Bee Pollen improves muscle protein and energy metabolism in malnourished old rats through interfering with the Mtor signaling pathway and mitochondrial activity. Nutrients 2014; 6(12): 5500-16.
[http://dx.doi.org/10.3390/nu6125500] [PMID: 25470375]

[59]   Abood SS, Ezat HN. Effect of adding different levels from Bee Pollen in diet on productive performance of broiler chickens. Plant Arch 2018; 18: 2435-8. [http://plantarchives.org/18-02/243--2438%20(4381).pdf].

[60]   Zeedan K, El-Neney BAM, Aboughaba AAAA, El-Kholy K. Efect of Bee Pollen at -different levels as natural additives on immunity and productive performance in rabbit males. Egyptian Poultry Science Journal 2017; 37: 213-31. [https://epsj.journals.ekb.eg/article_6778_4bebfa22d93a4c1e193f7a2807c8fef6.pdf].

[61]   Canogullar S, Baylan M, Sahinler N, Sahin A. Effects of propolis and pollen supplementations on growth performance and body components of Japanese quails *Coturnix japonica.* Arch Geflugelkd 2009; 73: 173-8.

[62]   Sevim B. Effects of supplemental Bee Pollen on performance, meat quality, serum constituents and immunity system in growing quails. S Afr J Anim Sci 2022; 51(6): 745-51.
[http://dx.doi.org/10.4314/sajas.v51i6.7]

[63]   Farag SA, El-Rayes TK. Effect of bee-pollen supplementation on performance, carcass traits and blood parameters of broiler chickens. Asian J Anim Vet Adv 2016; 11(3): 168-77.
[http://dx.doi.org/10.3923/ajava.2016.168.177]

[64]   Daudu OM. Bee Pollen extracts as potential antioxidants and inhibitors of α-amylase and α-glucosidase enzymes - *in vitro* assessment. J Apic Sci 2019; 63: 315-25.
[http://dx.doi.org/10.2478/jas-2019-0020]

[65]   Watford M. Small amounts of dietary fructose dramatically increase hepatic glucose uptake through a novel mechanism of glucokinase activation. Nutr Rev 2002; 60(8): 253-7.
[http://dx.doi.org/10.1301/002966402320289377] [PMID: 12199300]

[66]   Yang S, Qu Y, Chen J, *et al.* Bee Pollen polysaccharide from *Rosa rugosa* Thunb. (Rosaceae) promotes pancreatic β-cell proliferation and insulin secretion. Front Pharmacol 2021; 12: 688073.
[http://dx.doi.org/10.3389/fphar.2021.688073] [PMID: 34262457]

[67]   Rahayu AN, Bambang W, Merryana A, Soernarnatalina M, Dwi W, Sri H. Bee Pollen effect on blood glucose levels in alloxan-induced male wistar rats. Health Notions 2018; 2: 85.
[http://dx.doi.org/10.33846/hn.v2i1.85]

[68]   Setyawan AB, Satria AP, Arung ET, Paramita S. Effect of Bee Pollen kelulut bees on HBA1C in type 2 diabetes mellitus patients. Malays J Fundam Appl Sci 2023; 19(1): 56-60.
[http://dx.doi.org/10.11113/mjfas.v19n1.2750]

[69]   Panettieri V, Chatzifotis S, Messina CM, *et al.* Honey Bee Pollen in meagre (*Argyrosomus regius*) juvenile diets: effects on growth, diet digestibility, intestinal traits, and biochemical markers related to health and stress. Animals (Basel) 2020; 10(2): 231.
[http://dx.doi.org/10.3390/ani10020231] [PMID: 32023987]

[70]   Mohd Rosmi NSA, Shafie NH, Azlan A, Abdullah MA. Functional food mixtures: Inhibition of lipid peroxidation, HMGCoA reductase, and ACAT2 in hypercholesterolemia-induced rats. Food Sci Nutr 2021; 9(2): 875-87.
[http://dx.doi.org/10.1002/fsn3.2051] [PMID: 33598171]

[71]   El Ghouizi A, Bakour M, Laaroussi H, *et al.* Bee Pollen as functional food: insights into its composition and therapeutic properties. Antioxidants 2023; 12(3): 557.
[http://dx.doi.org/10.3390/antiox12030557] [PMID: 36978805]

[72]   Wang J, Jin GM, Zheng YM, Li SH, Wang H. Effect of Bee Pollen on development of immune organ of animal. Zhongguo Zhongyao Zazhi 2005; 30(19): 1532-6. [Chinese].
[PMID: 16335827]

CHAPTER 11

Propolis

Youssef A. Attia[1,2], **Mohamed E. Abd El-Hack**[3,*], **Mahmoud M. Alagawany**[3], **Salem R. Alyileili**[4], **Khalid A. Asiry**[2], **Saber S. Hassan**[1], **Asmaa Sh. Elnaggar**[1], **Hany I. Habiba**[2] and **Shatha I. Alqurashi**[5]

[1] *Animal and Poultry Production Department, Faculty of Agriculture, Damanhour University, Damanhour-22713, Egypt*

[2] *Sustainable Agriculture Production Research Group, Agriculture Department, Faculty of Environmental Sciences, King Abdulaziz University, Jeddah-20589, Saudi Arabia*

[3] *Poultry Department, Faculty of Agriculture, Zagazig University, Zagazig-44519, Egypt*

[4] *Department of Laboratory Analyses, College of food and Agriculture Sciences, United Arab Emirates University, AlAin United Arab Emirates*

[5] *Department of Biological Science, College of Science, University of Jeddah-21589, Jeddah, Saudi Arabia*

Abstract: Propolis, a resinous substance collected by bees from plant exudates and buds, has gained renewed interest as a natural feed additive for animal nutrition. Its composition varies depending on the plant source, time, and place of collection; however, it is primarily composed of phenolic acids, flavonoids, and their derivatives. Egyptian propolis contains phenolic acid esters (72.7%), dihydrochalcones (6.5%), flavones (4.6%), aliphatic acids (2.4%), flavanones (1.9%), chalcones (1.7%), phenolic acids (1.1%), and tetrahydrofuran (THF) derivatives (0.7%). Turkish propolis contains flavonoids (37.83%), organic acids (18.54%), aromatic acids and their esters (35.8%), hydrocarbons (4.89%), and other undefined components (2.94%). Propolis also contains vitamins, minerals, enzymes, fatty acids, amino acids, terpenes, and polysaccharides. Its bioactive components, including CAPE, artepillin C, caffeic acid, chrysin, galangin, quercetin, apigenin, kaempferol, pinobanksin, and pinocembrin contribute to its antibacterial, antifungal, anti-inflammatory, and antioxidant properties. These characteristics make propolis a promising natural growth promoter for livestock production and a potential replacement for antibiotics. In addition, propolis has applications in food technology as a preservative, with the added benefit of its residues being beneficial to human health. Ethanol is the preferred solvent for preparing propolis, although other solvents can also be used for the extraction and identification of its constituents. Polyphenols and flavonoids in propolis have been reported to positively affect the immune systems of various species, making it a valuable addition to livestock production practices.

* **Corresponding author Mohamed E. Abd El-Hack:** Poultry Department, Faculty of Agriculture, Zagazig University, Zagazig-44519, Egypt; E-mail: dr.mohamed.e.abdalhaq@gmail.com

Keywords: Animal performance, Antioxidants, Bioactive components, Growth promoters, Livestock, Propolis.

INTRODUCTION

The use of propolis as a feed supplement for animal nutrition has sparked interest in recent years. This material can be regarded as a natural source in the chemical and traditional medicine sectors [1]. Bees collect propolis, a raw resinous substance, from plant parts, exudates, and buds [2]. Bees use it to seal their hives [3] and, more importantly, to prevent the decomposition of animals that they have killed after invading the hive [4]. Propolis composition is influenced by several factors, including the source of the plant, time, and location of collection [5, 6].

Phenolic acids, phenolic acid esters, flavonoids, and terpenoids such as CAPE, artepillin C, caffeic acid, chrysin, galangin, quercetin, apigenin, kaempferol, pinobanksin 5-methyl ether, pinobanksin, pinocembrin, and pinobanksin 3-acetate are known to contribute to the biological activity of propolis [7, 8]. As of 2012, more than 500 composites have been found in propolis from various countries. They are classified as prenylated derivatives of coumarins, lignans, stilbenes, phenylpropanoids, terpenoids, and flavonoids [7, 9]. However, other typical chemical components, such as alkaloids and iridoids, have not been found in propolis, which is frequently attributed to plant sources [10 - 12].

Tetrahydrofuran derivatives (0.7%), phenolic acids (1.1%), chalcones (1.7%), flavanones (1.9%), aliphatic acids (2.4%), flavones (4.6%), dihydrochalcones (6.5%), and phenolic acid esters (72.7%) are the components of Egyptian propolis [13, 14]. Turkish propolis was found to contain 37.83% flavonoids, including pinocembrin, chrysin, and galangin; 18.54% organic acids; fatty acids, including n-hexadecanoic acid, coumaric acid, octadecanoic acid, cinnamic acid, and derivatives; 35.8% aromatic acids and their esters; 4.89% various hydrocarbons; and 2.94% other undefined components. These findings were also reported by Ozkok *et al.* [15]. Table **1** lists the elements of both the propolises.

Numerous vitamins (B1, B2, B6, C, and E), mineral elements (Ca, Cu, I, K, Mg, Na, Zn, Mn, and Fe), and enzymes (adenosine triphosphatase, succinic dehydrogenase, glucose-6-phosphatase, acid phosphatase, maltase, esterase, transhydrogenase α-amylase, β-amylase, α-lactamase, and β-lactamase) are all abundant in propolis [16 - 19]. Additionally, it contains several fatty acids, amino acids, terpenes, flavonoids, and derivatives of cinnamic acid [3, 20].

Vitamin A (6.1 IU/g of fresh matter and 8.1 IU/g of DM), vitamin B1 (4.5 µg/g on fresh matter basis and 6.5 µg/g of DM), vitamin B2 (20 µg/g of new matter and 28 µg/g of DM), and vitamin B6 (5 µg/g of fresh weight) were found to be present in propolis [21]. Cu (26.5 mg/kg) and Mn (40 mg/kg) were also found in propolis, according to this study, and the ash residue comprised Fe, Ca, Al, vanadium, strontium, and silicon.

Table 1. The bioactive components of the Egyptian and Turkish propolis (Ozkok *et al.* [15]).

Class of Components	%	References
Phenolic acid ester	72.7	Egyptian propolis Abd El-Hady and HegazI [13]
Phenolic acid	1.1	
Aliphatic acids	2.4	
Dihydrochalcones	6.5	
Chalcones	1.7	
Flavanones	1.9	
Flavones	4.6	
Tetrahydrofuran derivatives	0.7	
Flavonoids	37.83	Turkish propolis Ozkok *et al.* [3]
Organic acids and fatty acids	18.54	
Aromatic acids and their esters	35.8	
Various Hydrocarbon	4.89	
Other undefined components	2.94	

Atomic emission/absorption spectrometry was used to identify trace elements (Al, B, Ba, Cr, Fe, Mn, Ni, Sr, and Zn) and hazardous elements (As, Cd, Hg, and Pb) in propolis samples collected from several areas of Croatia [22]. Neutron activation analysis was used to identify many minerals, including Br, Co, Cr, Fe, Rb, Sb, Sm, and Zn, in several Argentinean propolis samples [23]. Based on their location, these investigations demonstrate that trace element profiles can be useful for propolis identification. Propolis frequently contains polysaccharides such as starch and the di- and monosaccharides fructose, ribose, rhamnose, talose, glucose, and saccharose [16, 24]. The most representative propolis compounds reported by Attia *et al.* [17] are listed in Table **2**.

Table 2. The major compounds in propolis (Attia *et al.* [17]).

Proximate Analysis of Propolis	%	Major Fatty Acids of Propolis	%
Crude protein	1	Palmitic	13.3
Ash	4.1	Stearic	6.4
Fat	1.2	Oleic	12.3
Carbohydrates	1.8	Linoleic	1.5

(Table 2) cont.....

Proximate Analysis of Propolis	%	Major Fatty Acids of Propolis	%
Essential oils	3.5	Linolenic	0.59
Flavonoids (Total)	25.9	Arachidonic	8

Bees collect propolis, a resinous substance converted into bee enzymes from plant exudates and buds [14]. Bees create it by combining plant exudates, beeswax, and salivary gland secretions; the hue ranges from green to dark brown [25]. Flavonoids and phenolic acids account for 40–70% of propolis, followed by waxes (20–35%), essential oils (1-3%), and approximately 5% of other organic materials [14, 26]. Propolis has also been linked to antibacterial, antifungal, anti-inflammatory, and antioxidant qualities [10, 11, 25].

More than 300 components are present in bee propolis, including flavonoids, amino acids, cinnamic and benzoic acids, and their esters, which replace phenolic acids and esters (Fig. **1**) [27]. Structural formulations of propolis-selected chemicals and the plant ecology of the region affect the qualitative and quantitative variations in the components of bee propolis [28].

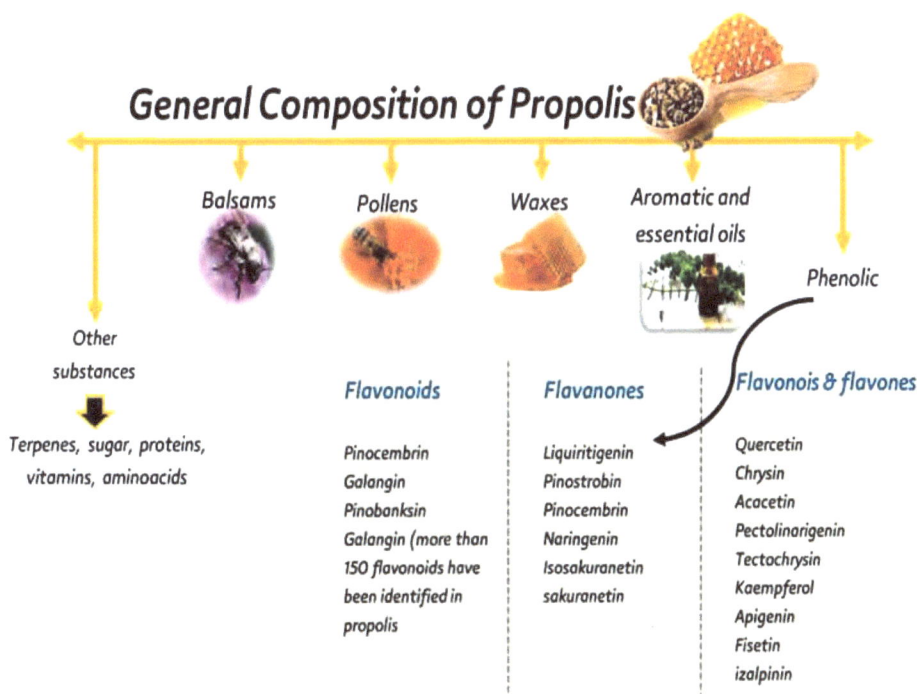

Fig. (1). General composition of propolis.

Because of this, some individuals use propolis as a general-purpose dietary supplement [6, 20, 29]. Products made from bees, particularly bee propolis, are very helpful in animal care and meat processing [6, 20, 25, 30]. Bee propolis is a popular "growth promoter" used by farmers to gain weight while using less feed. Food technology can benefit greatly from the antioxidant, antifungal, and antibacterial properties of bee propolis [10, 1, 31]. One unique feature of bee propolis, in contrast to other preservatives, is that its residues are good for human health [11, 18, 32]. The ideal solvent for propolis preparation is ethanol; however, water, ethyl ether, methanol, acetone, chloroform, and dichloromethane can be used to extract and identify propolis constituents [33].

Natural flavonoids contain polyphenols, flavonoids, phenolic acids, caffeic acid, and similar esters, and there are several publications that attest to the beneficial effects of flavonoids on the immune system (antibody production) of various species [34, 35].

Productive Performance

To make livestock production more profitable, it is important to increase the productivity and survival of the growing animals [8, 31, 36]. Antibiotics have been added to animal diets to promote growth and reduce the incidence of illnesses [13, 30]. However, careless use of antibiotics leads to the emergence of bacteria that are resistant to their effects, posing serious dangers to both human and animal health [25, 37]. Consequently, the use of antibiotics as growth promoters is forbidden in a number of nations, which poses a serious problem for those who produce meat and eggs [11, 17, 30]. These factors have led to a rise in the interest of researchers in discovering and creating novel natural substitutes for antibiotics and artificial antioxidants in recent years [31]. Natural alternatives include propolis produced by bees. Numerous bioactive metabolites with pharmacological qualities are present in these products [29, 38] (Fig. **2**).

Nonetheless, the benefits of propolis supplementation on livestock production have mostly been assessed in poultry [29, 39], and there is currently little data on the effects of propolis supplementation in cattle. It enhances livestock performance [28 - 30], lowers oxidative stress [24], and increases feed digestibility [40] in several domestic animal species, including sheep and broilers.

Studies have assessed the effects of propolis supplementation on productive performance, specifically in poultry [14, 20]. Furthermore, propolis supplementation has been shown in certain trials to be an efficient substitute for zinc bacitracin, the primary antibiotic used in rabbits, in rabbit diets without compromising animal health, mortality rates, or financial success [30, 41].

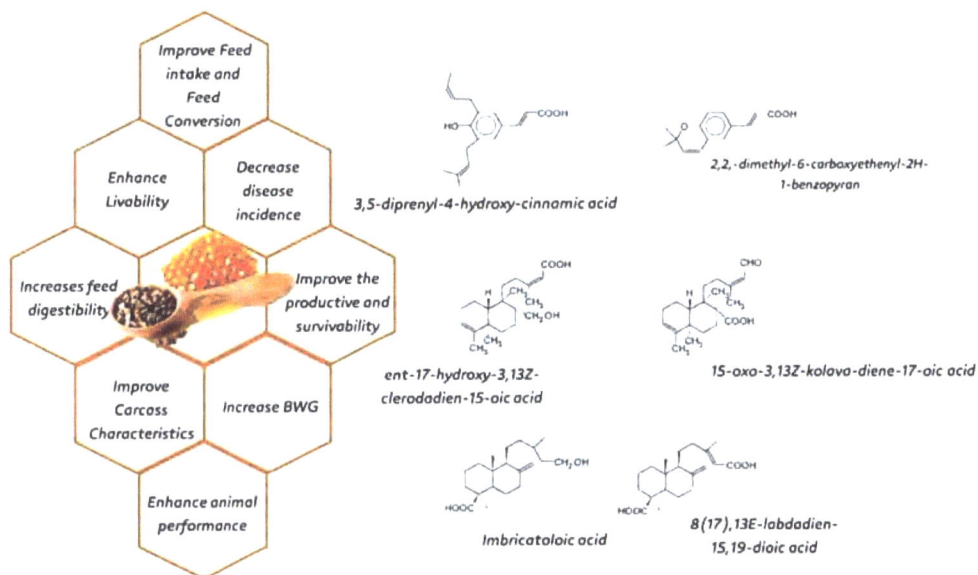

Fig. (2). Chemical structures of selected new propolis constituents and the beneficial effects on animal performance.

Research that has been released thus far [14, 17, 18, 20, 25, 30, 31] has demonstrated that propolis may be a viable substitute for improving the reproductive and production capacities as well as the general health of animals.

According to Abdel-Rahman and Mosaad [42], broiler ducks administered propolis at a feed rate of 2.0 g/kg gained weight at 12 weeks compared to the control group (7.393 and 6.242 kg, respectively). Likewise, feed conversion was better in the propolis-supplemented group (3.48 and 4.36 g feed/g gain, respectively) than in the control group. According to Abdel-Kareem and El-Sheikh [1], laying hens fed diets containing 250 and 1000 mg propolis/kg produced noticeably more eggs, both heavier and larger, than the control group. In comparison with the control group, the interior egg quality parameters increased significantly as propolis concentrations for supplemented hens increased, with the exception of yolk and albumen percentages. However, Ozkok *et al.* demonstrated that propolis doses of 0, 100, 200, and 400 mg/kg diet affected feed consumption, feed conversion ratio, egg production, egg weight, mortality, and egg quality parameters such as shell thickness and Haugh unit [15]. The laying hens' performance attributes, egg characteristics, and survival rate were not negatively affected by adding propolis to their food, according to the authors' conclusion. Different propolis dosages can be used to enhance egg production.

Propolis supplementation at 300 mg/kg diet administered continuously or intermittently increased the growth of broilers during the first 35 days of life, according to Attia *et al.* [17]. However, propolis supplementation at 200 mg/kg had no effect on the growth, total bacterial count, or mortality of NZW rabbits during the 35–94 day period [14]. At 90 days of age, the growth performance of the majority of the rabbits was unaffected by the propolis supplements in capsule form at doses of 150 mg and 300 mg. However, in comparison with the ZnB group, the growth rate was boosted by bee pollen 150 and bee pollen + propolis 300 [30].

Diets supplemented with 4 g/kg propolis extract increased BWG, specific growth rate, and feed intake reduction, and ultimately improved feed conversion and survival rate in Nile tilapia [44, 45]. However, other studies have shown that propolis has no effect on animal growth. For instance, propolis fortification did not increase the productivity of laying hens [45], pigs [46], fish [19, 47], developing rabbits [9], or ruminants [48, 49].

The results released thus far are neither consistent nor definitive. The variety of findings shown in fish [19] and animals supplemented with propolis [31] are linked to variations in dosage, length of the trial, supplementation techniques, and animal age. To create propolis-containing dietary supplements that can be utilized to enhance the health and productivity of rabbits, these sources of variability must be identified and managed.

Carcass Characteristics

Given that propolis has a complex nutritional composition that promotes both weight increase and muscle deposition, it is possible that animals fed this diet will have improved carcass features.

According to a meta-analysis by Sadarman *et al.*, propolis does not affect broiler carcass yield, abdominal fat, digestive organs, or visceral organs [29]. However, the weights of the spleen and breast meat increased linearly as propolis supplementation increased. Broiler diets containing propolis have been linked to increased bursa yield [50]. However, propolis's impact on the carcass characteristics of broilers was not confirmed by Arslan *et al.* [51] or Daza-Leon *et al.* [52].

Propolis administered at 150 and 200 mg/kg resulted in improved carcass production, a greater proportion of edible portions and spleen, and a decreased percentage of abdominal fat in rabbits compared with the control group [53]. Attia *et al.* [14] and Sierra-Galícia *et al.* [18] reported improved carcass features in rabbits fed with 200 mg/kg BW and 50 µL/kg BW, respectively. Furthermore, propolis supplements may lower the amount of fat in the abdomen [36].

However, in experiments involving lambs, Silva *et al.* [54] observed deterioration in hot and cold dressing as a result of brown propolis supplementation. Numerous studies have found no difference in the carcass characteristics of animals given propolis-containing diets, including lambs [57], quails [58], broilers [50, 59], and bulls [55, 56].

Biochemical Contents and Antioxidant Status

Because it inhibits intestinal maltase activity, modifies postprandial blood glucose levels, increases glucose consumption by peripheral tissues, and activates the hepatic enzyme glucokinase, propolis exhibits antihyperglycemic effects [60, 61]. Propolis also exhibits hypolipidemic action, lowering blood levels of LDL-c, total triglycerides, and cholesterol while enhancing hepatic and renal functions, as well as antioxidant defenses [61, 62].

Studies on the effects of propolis on the blood antioxidant status [8, 36] and blood biochemistry [9, 17, 18], specifically in poultry, have been conducted. Propolis may mitigate the negative effects of high stocking density on quail antioxidant enzymes and lipid peroxidation, as demonstrated by Arslan *et al.* [63]. Zeweill *et al.* [58] found that Japanese quails treated with propolis or propolis + ginger had higher levels of glutathione peroxidase, humoral immunity, total antioxidant capacity, and HDL-c and lower levels of total lipids, triglycerides, total cholesterol, and LDL-c.

The high stocking quail group's blood MDA levels were found to be substantially higher (6.84 nM/mL) than those of the groups supplementing with 0.5 g (4.33 nM/mL), 1.0 g (4.08 nM/mL), and 1.5 g (4.56 nM/mL) of propolis/kg, according to Arslan and Seven [63]. The impact was ascribed to the phenolic acid, terpenoid, and flavonoid concentrations of propolis, which have antioxidative properties.

When propolis was fortfieded to the meals of laying quails that were subjected to heat stress, the eggs produced had a greater yolk index, egg weight, and eggshell thickness. In addition, body temperature, corticosterone, malonaldehyde, ALT, and total cholesterol levels were all lower in the birds than in the control group [64].

Hassan *et al.* [59] showed improvements in the levels of total protein, triglycerides, and total cholesterol as a result of the broiler feed supplementation with 1, 2, and 3 g/kg. Propolis (50, 100, 200, and 300 ppm) addition in the broiler diet did not, however, appear to have any effect on biochemical parameters or antibody titers against the Newcastle and influenza viruses, according to Gheisari *et al.* [65].

Propolis has been found in studies involving rabbits to have a promising effect on lowering blood lipid levels, including total lipid [14], total cholesterol, triglycerides [53], and LDL-c [36], as well as enhancing humoral and cellular immunity [9]. Waly *et al.* [53] also observed that the addition of 100, 150, and 200 mg/kg propolis diets resulted in a decrease in malondialdehyde levels. Nevertheless, the blood serum biochemistry of propolis-supplemented rabbits showed no alterations, according to recent studies by Gabr *et al.* [66], Al-Homidan *et al.* [9], and Sierra-Galicia *et al.* [18]. Increased plasma levels of total protein, albumin, globulin, glucose, and total lipids were observed in rabbits fed a 200 mg/kg diet containing propolis; however, decreased levels of total cholesterol, urea, urea/creatinine ratio, AST, ALT, and AST/ALT ratio were observed [20]. Abdel-Kareem and El-Sheikh [1] found that while cholesterol and liver enzymes were significantly decreased, total protein and globulin concentrations rose as propolis level concentration increased (0 *vs.* 250 and 1000 mg/kg diet).

Propolis (150 mg/kg) was shown to lower plasmatic levels of T3 in comparison to the control group by Attia *et al.* [41]. Plasma AST levels were lower in the propolis-fed group (300 mg/kg) than in the propolis-fed group (150 mg/kg).

Propolis supplementation has also been studied in fish. The total protein and globulin levels in the blood of Nile tilapia were elevated when they were fed 0.4% ethanolic extract of propolis [44]. Hassaan *et al.* [43] reported reduced activity of ALT, AST, triglycerides, and total cholesterol compared to the control group for tilapias submitted to cold stress; this indicated the hepatoprotective and cardiac benefits of propolis.

Ruminants frequently suffer from heat stress in a variety of production settings, which alters their respiratory and metabolic rates [67]. By increasing the levels of antioxidant enzymes and overall antioxidant capacity, and boosting the immune system, propolis may help ruminants counteract the negative effects of heat stress [68, 69]. The addition of 3 g of red propolis extract/ewe/day for 45 days was assessed by Morsy *et al.* [40], who found improvements in the blood levels of cortisol, globulin, and total protein. However, the glucose level increased.

Haematological Parameters

Propolis has an immunostimulatory effect and enhances the efficiency of hemoglobin regeneration and iron consumption [10, 11]. Additionally, it helps shield blood cells from lipid peroxidation because of its antioxidant properties.

In comparison to laying hens fed diets without propolis or with 0.5 and 1 g/kg of propolis, the use of 3 g/kg of propolis enhanced erythrocyte count; hematocrit, hemoglobin, and leukocyte number were unaffected [70]. Mehaisen *et al.* [64] observed that Japanese quail fed 1 g/kg of propolis had a decrease in heterophils/lymphocytes and an increase in white blood cells.

According to Attia *et al.*, propolis supplementation of 250 mg/kg feed in hens resulted in enhanced red blood cells, hemoglobin levels, and lymphocyte count [17]. According to Shaddel-Tili *et al.* [71], broiler chickens given a dosage of 2.0 g propolis/kg food had significantly higher packed cell volume and heterophil counts. Supplementing chicken diets with propolis may help avoid anemia and enhance haematological characteristics [6, 10, 11, 41, 72].

Nonetheless, according to Abbas [73], propolis (0–2.5 g/kg) had no influence on the hematocrit and hemoglobin levels of broilers at 28 days of age. Hassan *et al.* [59] observed the same thing in broilers given diets containing 1, 2, and 3 g/kg of propolis. Hemoglobin levels, blood cell counts, and the proportion of monocytes, lymphocytes, and heterophils were all the same according to the authors.

Research by Shahin *et al.* with Nile tilapia [44] revealed improvements in leukocyte counts, the proportion of neutrophils, lymphocytes, monocytes, and phagocytic index when compared to fish fed diets devoid of propolis. In addition, Muzzolon *et al.* [47] observed that young pacu fortfed with 3% propolis ethanolic extract for 60 days had higher platelet and neutrophil levels than the control group.

The use of propolis in rabbits has been assessed by Attia *et al.* (2013), Hashem *et al.* [36], and Sierra-Galicia *et al.* [18]. When Slanzon *et al.* [49] investigated the effects of red propolis extract in calves, they found no changes in the levels of hemoglobin, hematocrit, or blood cell count.

Immunity and Health

Propolis is effective against several viruses [76, 77], fungi [78, 79], and bacteria [74, 75]. Its non-specific immunostimulatory properties have also been demonstrated [47]. The impact of some bee products on the immunological response of hens infected with virulent Newcastle disease virus (NDV) was investigated by Hegazi *et al.* [34]. Compared to infected mice without supplementation, those infected with virulent NDV and later supplemented with propolis or honey had a higher survival rate.

The phenolic acid, flavonoid, and terpenoid concentrations of propolis are the basis for its antioxidative, cytostatic, antimutagenic, and immunomodulatory properties [80, 81]. Flavonoids, which are abundant secondary metabolites found in plants, are interesting natural products [82].

Numerous findings have attested to the beneficial benefits of naturally occurring flavonoids in various animal immune systems. Many aspects have been taken into consideration in this association, with the main emphasis of these investigations being antibody production, T lymphocyte stimulation, boosting blood lymphocytes, phagocytosis activity, and thymus and bursa of Fabricius weight [34, 35]. Several studies have linked propolis to immunostimulatory effects and overall health maintenance [83, 84]. The addition of PAE did not enhance the immune system of broilers. Propolis supplements have been shown to enhance immunological responses [12]. Several parameters are considered in this relationship, including the bursa of Fabricius weight, phagocytosis activity, T lymphocyte activation, and an increase in blood lymphocytes [34, 35]. The cytokines generated by T cells that are activated and stimulated by ethanol propolis extract are linked to both humoral and cellular immune responses [85]. Propolis and its extracts have been shown in numerous verified studies to activate the immune system in both humans and mice. This activation includes increased levels of IL-1 [83, 86, 87], IL-2 [86, 88], IL-4 [88], antibody response [85, 88], T lymphocyte proliferation, an increase in the ratio of CD4+ to CD8+, and macrophage activation [10, 88 - 90].

Propolis may directly influence the digestive system by activating lymphatic tissue and indirectly by altering the bacterial population in the GIT lumen [91]. In this regard, combining these reactions may be associated with an increased humoral response in broilers. Propolis exhibits antiinflammatory [90] and antioxidant properties [92 - 94]. It is thought that this is because it inhibits prostaglandin formation [95], which is an anti-immune chemical that improves humoral responses.

Propolis or its extracts have been shown in several studies to have positive effects on immune system activation in both humans and animals. These effects include increased CD4+ /CD8+ ratio, T-cell proliferation, and macrophage activation [88 - 90]. Propolis did not alter the antibody titer against the NDV vaccination, as demonstrated by Kiaei *et al.* [96]. It should be noted that the response to forced vaccination may be significantly higher than that of influenza antibodies produced in response to spontaneous infection by environmental serotypes. According to Taheri *et al.* [91], supplementation with oil-extracted propolis (OEP) boosted antibody titers against infectious bursal disease, Newcastle disease, and avian influenza.

The impact of OEP (0, 40, 70, 100, 400, 700, and 1000 mg/kg diet) on the immune system of broilers aged 1–47 days was studied by Ziaran *et al.* [97] and Haščík *et al.* [98]. They discovered that hens fed with 70 and 100 mg/kg of OEP had an enhanced antibody response to Newcastle disease at 30 days of age or 20 days after vaccination. Histological examination showed that hens administered 1000 mg of OEP daily had more proliferative cells in the bursa of the Fabricius.

Leukocyte counts in the intestinal lamina propria of broilers fed diets containing high doses of OEP (400, 700, and 1000 mg) were higher than those of birds fed diets containing lower amounts of OEP. At the maximum dosage (1000 mg) of OEP, there were more lymphoid cells in the liver portal region, and the control therapy resulted in a thicker blood vessel wall.

According to Eyng *et al.* [99], broiler feed with 100 mg/kg crude propolis was found to be an efficient immunostimulant for cell-mediated responses in the initial phase. Babaei *et al.* [100] compared the propolis (1.0 g/kg) in quail food to the control and antibiotic (virginiamycin) groups and observed an increase in the Newcastle disease titer. High doses of propolis may modify humoral immunity in broilers, as seen by the considerably greater serum IgG and IgM concentrations in birds given 0.7, 0.8, and 0.9 g/kg of propolis compared to control chickens [101]. The type and amount of propolis and its ingredients in the diets, length of feeding with experimental diets, research population (*i.e.*, age, weight, or breed), and animal species may all have an impact on discrepancies in the results of various studies.

CONCLUSION AND REMARKS

In conclusion, propolis shows significant potential as a natural feed additive for animal nutrition because of its diverse bioactive components and beneficial properties. Its antibacterial, antifungal, anti-inflammatory, and antioxidant effects make it a promising alternative to antibiotics as a growth promoter in livestock production. Studies have demonstrated that up to 1000 mg/kg of propolis can positively affect productive performance, carcass characteristics, biochemical parameters, antioxidant status, hematology, and immunity across various animal species when fed at appropriate levels. However, the results have varied across studies, likely due to variations in propolis composition, dosage, duration of supplementation, and animal factors. Although propolis shows promise, further research is needed to optimize its use and fully understand its mechanisms of action in different livestock species. Overall, propolis represents an exciting natural option to enhance animal health, performance, and product quality, such as shelf life, because the livestock industry avoids antibiotic growth promoters. Continued investigation of the standardization and application of propolis in animal feeds is warranted to realize its full potential as a sustainable feed additive.

IMPLICATIONS

These findings on propolis as a natural feed additive have significant applications in the livestock industry. As concerns over antibiotic resistance grow and regulations on antibiotic use in animal production tighten, propolis offers a promising alternative to maintain animal health, product quality, and performance. Their diverse bioactive components and beneficial properties could potentially replace or reduce the need for synthetic growth promoters and antibiotics. This shift towards natural additives, such as propolis, aligns with consumer demands for more sustainable and antibiotic-free animal products. However, inconsistent results across studies highlight the need for further research to standardize propolis composition and optimize its use in different animal species and production systems. The successful implementation of propolis in animal feed could lead to improved animal welfare, enhanced product quality, and more sustainable livestock production practices. Additionally, the potential human health benefits of consuming animal products with propolis residues could have broader implications for public health. As the livestock industry continues to evolve, propolis may play a crucial role in the transition towards more natural and sustainable production methods.

REFERENCES

[1]   Abdel-Kareem AAA, El-Sheikh TM. Impact of supplementing diets with propolis on productive performance, egg quality traits and some haematological variables of laying hens. J Anim Physiol Anim Nutr (Berl) 2017; 101(3): 441-8.
[http://dx.doi.org/10.1111/jpn.12407] [PMID: 26614568]

[2]   Bobiş O. Plants: sources of diversity in propolis properties. Plants 2022; 11(17): 2298.
[http://dx.doi.org/10.3390/plants11172298] [PMID: 36079680]

[3]   Wagh VD. Propolis: a wonder bees product and its pharmacological potentials. Adv Pharmacol Sci 2013; 2013: 1-11.
[http://dx.doi.org/10.1155/2013/308249] [PMID: 24382957]

[4]   Brumfitt W, Hamilton-Miller JM, Franklin I. Antibiotic activity of natural products: 1. Propolis. Microbios 1990; 62(250): 19-22.
[PMID: 2110610]

[5]   Markham KR, Mitchell KA, Wilkins AL, Daldy JA, Lu Y. HPLC and GC-MS identification of the major organic constituents in New Zeland propolis. Phytochemistry 1996; 42(1): 205-11.
[http://dx.doi.org/10.1016/0031-9422(96)83286-9]

[6]   Attia YA, Al-Khalaifah H, Ibrahim MS, Al-Hamid AEA, Al-Harthi MA, Elnaggar Sh. El-Naggar ShA. Blood haematological and biochemical constituents, antioxidant enzymes, immunity and lymphoid

organs of broiler chicks supplemented with propolis, bee pollen and mannan oligosaccharides continuously or intermittently. Poult Sci 2017; 96(12): 4182-92.
[http://dx.doi.org/10.3382/ps/pex173] [PMID: 29053876]

[7]   Huang S, Zhang CP, Wang K, Li G, Hu FL. Recent advances in the chemical composition of propolis. Molecules 2014; 19(12): 19610-32.
[http://dx.doi.org/10.3390/molecules191219610] [PMID: 25432012]

[8]   Hassan SS, Shahba HA, Mansour M. Influence of using date palm pollen or bee pollen on some blood biochemical metabolites, semen characteristics and subsequent reproductive performance of v-line male rabbits. Egyptian Journal of Rabbit Science 2022; 32(1): 19-39.
[http://dx.doi.org/10.21608/ejrs.2022.232781]

[9]   Al-Homidan I, Fathi M, Abdelsalam M, *et al.* Effect of propolis supplementation and breed on growth performance, immunity, blood parameters and cecal microbiota in growing rabbits. Animal Bioscience 2022; 35(10): 1606-15.
[http://dx.doi.org/10.5713/ab.21.0535] [PMID: 35507863]

[10]   Alagawany M, Qattan SYA, Attia YA, *et al.* Use of chemical nano-selenium as an antibacterial and antifungal agent in quail diets and its effect on growth, carcasses, antioxidant, immunity and caecal microbes. Animals (Basel) 2021; 11(11): 3027.
[http://dx.doi.org/10.3390/ani11113027] [PMID: 34827760]

[11]   Attia YA, Giorgio GM, Addeo NF, *et al.* COVID-19 pandemic: impacts on bees, beekeeping, and potential role of bee products as antiviral agents and immune enhancers. Environ Sci Pollut Res Int 2022; 29(7): 9592-605.
[http://dx.doi.org/10.1007/s11356-021-17643-8] [PMID: 34993785]

[12]   AL-Kahtani SN, Alaqil AA, Abbas AO. Modulation of antioxidant defense, immune response, and growth performance by inclusion of propolis and bee pollen into broiler diets. Animals (Basel) 2022; 12(13): 1658.
[http://dx.doi.org/10.3390/ani12131658] [PMID: 35804557]

[13]   Abd El-Hady FK, Hegazi AG. Gas chromatography – mass spectrometry (GC/MS) study of the Egyptian propolis 1- aliphatic, phenolic acids and their esters. Egyptian Journal of Applied Science 1994; 9: 749-60.

[14]   Attia YA, El-Hanoun AM, Bovera F, Monastra G, El-Tahawy WS, Habiba HI. Growth performance, carcass quality, biochemical and haematological traits and immune response of growing rabbits as affected by different growth promoters. J Anim Physiol Anim Nutr (Berl) 2014; 98(1): 128-39.
[http://dx.doi.org/10.1111/jpn.12056] [PMID: 23419029]

[15]   Ozkok D, Iscan KM, Ilici S. Effects of dietary propolis supplementation on performance and egg quality in laying hens. J Anim Vet Adv 2013; 12: 269-75.
[http://dx.doi.org/10.36478/javaa.2013.269.275]

[16]   Kurek-Górecka A, Rzepecka-Stojko A, Górecki M, Stojko J, Sosada M, Świerczek-Zięba G. Structure and antioxidant activity of polyphenols derived from propolis. Molecules 2013; 19(1): 78-101.
[http://dx.doi.org/10.3390/molecules19010078] [PMID: 24362627]

[17]   Attia YA, Al-Hamid AEA, Ibrahim MS, Al-Harthi MA, Bovera F, Elnaggar Sh. Productive performance, biochemical and hematological traits of broiler chickens supplemented with propolis, bee pollen, and mannan oligosaccharides continuously or intermittently. Livest Sci 2014; 164: 87-95.
[http://dx.doi.org/10.1016/j.livsci.2014.03.005]

[18]   Sierra-Galicia MI, Rodríguez-de Lara R, Orzuna-Orzuna JF, *et al.* Supplying bee pollen and propolis to growing rabbits: effects on growth performance, blood metabolites, and meat quality. Life (Basel) 2022; 12(12): 1987.
[http://dx.doi.org/10.3390/life12121987] [PMID: 36556352]

[19]   Santos EL, Barbosa JM, Porto-Neto FF, *et al.* Propolis extract as a feed additive of the Nile tilapia juveniles. Arq Bras Med Vet Zootec 2023; 75(4): 744-52.

[http://dx.doi.org/10.1590/1678-4162-12806]

[20]   Attia YA, Bovera F, El-Tahawy WS, El-Hanoun AM, Al-Harthi MA, Habiba HI. Productive and reproductive performance of rabbits does as affected by bee pollen and/or propolis, inulin and/or mannan-oligosaccharides. World Rabbit Sci 2015; 23(4): 273-82.
[http://dx.doi.org/10.4995/wrs.2015.3644]

[21]   Moreira TF. Composição química da própolis: vitaminas e aminoácidos. Rev Bras Farmacogn 1986; 1(1): 12-9.
[http://dx.doi.org/10.1590/S0102-695X1986000100003]

[22]   Cvek J, Medić-Šarić M, Vitali D, Vedrina-Dragojević I, Šmit Z, Tomić S. The content of essential and toxic elements in Croatian propolis samples and their tinctures. J Apic Res 2008; 47(1): 35-45.
[http://dx.doi.org/10.1080/00218839.2008.11101421]

[23]   Cantarelli MA, Camiña JM, Pettenati EM, Marchevsky EJ, Pellerano RG. Trace mineral content of Argentinean raw propolis by neutron activation analysis (NAA): assessment of geographical provenance by chemometrics. Food Sci Technol (Campinas) 2011; 44: 256-60.
[http://dx.doi.org/10.1016/j.lwt.2010.06.031]

[24]   Cécere BGO, da Silva AS, Molosse VL, *et al.* Addition of propolis to milk improves lactating lamb's growth: Effect on antimicrobial, antioxidant and immune responses in animals. Small Rumin Res 2021; 194: 106265.
[http://dx.doi.org/10.1016/j.smallrumres.2020.106265]

[25]   Hashem NM, Hassanein EM, Simal-Gandara J. Simal-GandaraJ. Improving reproductive performance and health of mammals using honeybee products. Antioxidants 2021; 10(3): 336.
[http://dx.doi.org/10.3390/antiox10030336] [PMID: 33668287]

[26]   Osés SM, Marcos P, Azofra P, de Pablo A, Fernández-Muíño MÁ, Sancho MT. Phenolic profile, antioxidant capacities and enzymatic inhibitory activities of propolis from different geographical areas: Needs for analytical harmonization. Antioxidants 2020; 9(1): 75.
[http://dx.doi.org/10.3390/antiox9010075] [PMID: 31952253]

[27]   Bankova VS, de Castro SL, Marcucci MC. Propolis: recent advances in chemistry and plant origin. Apidologie (Celle) 2000; 31(1): 3-15.
[http://dx.doi.org/10.1051/apido:2000102]

[28]   Martinello M, Mutinelli F. Antioxidant activity in bee products: a review. Antioxidants 2021; 10(1): 71.
[http://dx.doi.org/10.3390/antiox10010071] [PMID: 33430511]

[29]   Sadarman S, Erwan E, Irawan A, *et al.* Propolis supplementation affects performance, intestinal morphology, and bacterial population of broiler chickens. S Afr J Anim Sci 2021; 51(4): 477-87.
[http://dx.doi.org/10.4314/sajas.v51i4.8]

[30]   Attia YA, Bovera F, Abd-Elhamid AEHE, *et al.* Evaluation of the carryover effect of antibiotic, bee pollen and propolis on growth performance, carcass traits and splenic and hepatic histology of growing rabbits. J Anim Physiol Anim Nutr (Berl) 2019; 103(3): 947-58.
[http://dx.doi.org/10.1111/jpn.13068] [PMID: 30714248]

[31]   Sierra-Galicia MI, Rodríguez-de Lara R, Orzuna-Orzuna JF, Lara-Bueno A, Ramírez-Valverde R, Fallas-López M. Effects of supplementation with bee pollen and propolis on growth performance and serum metabolites of rabbits: a meta-analysis. Animals (Basel) 2023; 13(3): 439.
[http://dx.doi.org/10.3390/ani13030439] [PMID: 36766327]

[32]   Kamel KI, El-Hanoun AM, El-Sbeiy MS, Gad HAM. Effect of bee propolis extract (bee glue) on some productive, reproductive and physiological traits of rabbits does and their progenys. 5th International Conference on Rabbit Production in Hot Climate. Hurghada, Egypt. 2007; pp. 403-15.

[33]   Szliszka E, Kucharska AZ. Sokol- Letiwska A, Mertas A, Czuba ZP, Krol W. Chemical composition and antiinflammatory effect of ethanolic extract of Brazilian green propolis on activated J774A.1

Macrophages. Evid Based Complement Alternat Med 2013; 2013: 976415.
[http://dx.doi.org/10.1155/2013/976415] [PMID: 23840273]

[34] Hegazi AG, El Miniawy HF, ElMiniawy FA. Effect of some honeybee products on immune response of chicken infected with Virulent NDV. Egypt J Immunol 1995; 2: 79-86.

[35] Kong X, Hu Y, Rui R, Wang D, Li X. Effects of Chinese herbal medicinal ingredients on peripheral lymphocyte proliferation and serum antibody titer after vaccination in chicken. Int Immunopharmacol 2004; 4(7): 975-82.
[http://dx.doi.org/10.1016/j.intimp.2004.03.008] [PMID: 15233143]

[36] Hashem NM, Abd El-Hady AM, Hassan OA. Inclusion of phytogenic feed additives comparable to vitamin E in diet of growing rabbits: Effects on metabolism and growth. Ann Agric Sci 2017; 62(2): 161-7.
[http://dx.doi.org/10.1016/j.aoas.2017.11.003]

[37] Huang P, Cui X, Wang Z, *et al.* Effects of *Clostridium butyricum* and a bacteriophage cocktail on growth performance, serum biochemistry, digestive enzyme activities, intestinal morphology, immune responses, and the intestinal microbiota in rabbits. Antibiotics (Basel) 2021; 10(11): 1347.
[http://dx.doi.org/10.3390/antibiotics10111347] [PMID: 34827285]

[38] Abdelnour SA, Abd El-Hack ME, Alagawany M, Farag MR, Elnesr SS. Beneficial impacts of bee pollen in animal production, reproduction and health. J Anim Physiol Anim Nutr (Berl) 2019; 103(2): 477-84.
[http://dx.doi.org/10.1111/jpn.13049] [PMID: 30593700]

[39] Lika E, Kostić M, Vještica S, Milojević I, Puvača N. Honeybee and plant products as natural antimicrobials in enhancement of poultry health and production. Sustainability (Basel) 2021; 13(15): 8467.
[http://dx.doi.org/10.3390/su13158467]

[40] Morsy AS, Soltan YA, El-Zaiat HM, Alencar SM, Abdalla AL. Bee propolis extract as a phytogenic feed additive to enhance diet digestibility, rumen microbial biosynthesis, mitigating methane formation and health status of late pregnant ewes. Anim Feed Sci Technol 2021; 273: 114834.
[http://dx.doi.org/10.1016/j.anifeedsci.2021.114834]

[41] Attia YA, Bovera F, Abd Elhamid AEH, *et al.* Bee pollen and propolis as dietary supplements for rabbit: Effect on reproductive performance of does and on immunological response of does and their offspring. J Anim Physiol Anim Nutr (Berl) 2019; 103(3): 959-68.
[http://dx.doi.org/10.1111/jpn.13069] [PMID: 30714649]

[42] Abdel-Rahman MA, Mosaad GM. Effect of propolis as additive on some behavioural patterns, performance and blood parameters in muscovy broiler ducks. J Adv Vet Res 2013; 3: 64-8.

[43] Hassaan MS, EL Nagar AG, Salim HS, Fitzsimmons K, El-Haroun ER. Nutritional mitigation of winter thermal stress in Nile tilapia by propolis-extract: Associated indicators of nutritional status, physiological responses and transcriptional response of delta-9-desaturase gene. Aquaculture 2019; 511: 734256.
[http://dx.doi.org/10.1016/j.aquaculture.2019.734256]

[44] Shahin S, Eleraky W, Elgamal M, Hassanein E, Ibrahim D. Effect of olive leaves and propolis extracts on growth performance, immunological parameters and economic efficiency using Nile tilapia (*Oreochromis niloticus*). Zagazig Vet J 2019; 47(4): 447-58.
[http://dx.doi.org/10.21608/zvjz.2019.17129.1085]

[45] Galal A, Abd El - M AM, Ahmed AMH, Zaki TG. Abd EL-Motaal AM, Ahmed AMH, Zaki TG. Productive performance and immune response of laying hens as affected by dietary propolis supplementation. Int J Poult Sci 2008; 7(3): 272-8.
[http://dx.doi.org/10.3923/ijps.2008.272.278]

[46] Gonçalves LMP, Kiefer C, Silva CM, *et al.* Propolis extract in the diet of weaned piglets. Cienc Rural 2018; 48: e20161056.

[http://dx.doi.org/10.1590/0103-8478cr20161056]

[47]   Muzzolon A, Bicudo ÁJA, Oldoni TLC, Sado RY. Dietary brown propolis extract modulated non-specific immune system and intestinal morphology of Pacu *Piaractus mesopotamicus*. Braz Arch Biol Technol 2021; 64: e21200787.
[http://dx.doi.org/10.1590/1678-4324-2021200787]

[48]   Silva JA, Ítavo CCBF, Ítavo LCV, *et al.* Effects of dietary brown propolis on nutrient intake and digestibility in feedlot lambs. Rev Bras Zootec 2014; 43(7): 376-81.
[http://dx.doi.org/10.1590/S1516-35982014000700006]

[49]   Slanzon GS, Toledo AF, Silva AP, *et al.* Red propolis as an additive for preweaned dairy calves: Effect on growth performance, health, and selected blood parameters. J Dairy Sci 2019; 102(10): 8952-62.
[http://dx.doi.org/10.3168/jds.2019-16646] [PMID: 31421873]

[50]   Alani AAT, Alheeti ASA, Alani ENS. Comparison between effect of adding propolis and antibiotic in broiler chickens on productive performance and carcass traits. International Conference on Agricultural Sciences 2019; 388: 012032.
[http://dx.doi.org/10.1088/1755-1315/388/1/012032]

[51]   Arslan AS, Seven PT, Yilmaz S, Seven I. The effects of propolis on performance, carcass and antioxidant status characteristics in quails reared under different stocking density. European Poultry Science 2014; 78: eps 2014. 20.
[http://dx.doi.org/10.1399/eps.2014.20]

[52]   Daza-Leon C, Gomez AP, Álvarez-Mira D, *et al.* Characterization and evaluation of Colombian propolis on the intestinal integrity of broilers. Poult Sci 2022; 101(12): 102159.
[http://dx.doi.org/10.1016/j.psj.2022.102159] [PMID: 36279608]

[53]   Waly A, Abo El-Azayem E, Younan G, Zedan A, El- Komy H, Mohamed R. Effects of propolis supplementation on growth performance, nutrients digestibility, carcass characteristics and meat quality of growing New Zealand rabbits. Egypt J Nutr Feeds 2021; 24(2): 65-73.
[http://dx.doi.org/10.21608/ejnf.2021.210779]

[54]   da Silva J, Ítavo CC, Ítavo LC, *et al.* Dietary addition of crude form or ethanol extract of brown propolis as nutritional additive on behaviour, productive performance and carcass traits of lambs in feedlot. J Anim Feed Sci 2019; 28(1): 31-40.
[http://dx.doi.org/10.22358/jafs/105442/2019]

[55]   Aguiar SC, Zeoula LM, Moura LPP, Prado IN, Paula EM, Samensari RB. Performance, digestibility, microbial production and carcass characteristics of feedlot young bulls fed diets containing propolis. Acta Sci Anim Sci 2012; 34(4): 393-400.
[http://dx.doi.org/10.4025/actascianimsci.v34i4.15167]

[56]   Valero MV, Zeoula LM, Moura LPP, Costa Júnior JBG, Sestari BB, Prado IN. Propolis extract in the diet of crossbred (½ Angus *vs.* ½ Nellore) bulls finished in feedlot: animal performance, feed efficiency and carcass characteristics. Semin Cienc Agrar 2015; 36(2): 1067-78.
[http://dx.doi.org/10.5433/1679-0359.2015v36n2p1067]

[57]   Ítavo CC, Ítavo LC, Esteves CA, *et al.* Influence of solid residue from alcoholic extraction of brown propolis on intake, digestibility, performance, carcass and meat characteristics of lambs in feedlot. J Anim Feed Sci 2019; 28(2): 149-58.
[http://dx.doi.org/10.22358/jafs/109284/2019]

[58]   Zeweil HS, Zahran SM, Abd El-Rahman MHA, Dosoky WM, Abu Hafsa SH, Moktar AA. Effect of using bee propolis as natural supplement on productive and physiological performance of Japanese quail. Egypt Poult Sci 2016; 36(1): 161-75.
[http://dx.doi.org/10.21608/epsj.2016.33248]

[59]   Hassan RIM, Mosaad GMM, Abd El-Wahab HY. Effect of feeding propolis on growth performance of broilers. J Adv Vet Res 2018; 8: 66-72.

[60]   Al-Hariri M. Propolis and its direct and indirect hypoglycemic effect. J Family Community Med 2011; 18(3): 152-4.
[http://dx.doi.org/10.4103/2230-8229.90015] [PMID: 22175043]

[61]   Oršolić N, Landeka Jurčević I, Đikić D, *et al.* Effect of propolis on diet-induced hyperlipidemia and atherogenic indices in mice. Antioxidants 2019; 8(6): 156.
[http://dx.doi.org/10.3390/antiox8060156] [PMID: 31163593]

[62]   Silva DB, Miranda AP, Silva DB, *et al.* Propolis and swimming in the prevention of atherogenesis and left ventricular hypertrophy in hypercholesterolemic mice. Braz J Biol 2015; 75(2): 414-22.
[http://dx.doi.org/10.1590/1519-6984.15313] [PMID: 26132026]

[63]   Sur Arslan A, Tatlı Seven P. The effects of propolis on serum malondialdehyde, fatty acids and some blood parameters in Japanese quail (*Coturnix coturnix japonica*) under high stocking density. J Appl Anim Res 2017; 45(1): 417-22.
[http://dx.doi.org/10.1080/09712119.2016.1206901]

[64]   Mehaisen GMK, Desoky AA, Sakr OG, Sallam W, Abass AO. Propolis alleviates the negative effects of heat stress on egg production, egg quality, physiological and immunological aspects of laying Japanese quail. PLoS One 2019; 14(4): e0214839.
[http://dx.doi.org/10.1371/journal.pone.0214839] [PMID: 30964896]

[65]   Gheisari A, Shahrvand S, Landy N. Effect of ethanolic extract of propolis as an alternative to antibiotics as a growth promoter on broiler performance, serum biochemistry, and immune responses. Vet World 2017; 10(2): 249-54.
[http://dx.doi.org/10.14202/vetworld.2017.249-254] [PMID: 28344411]

[66]   Gabr S, Younan G, Hamad M, Ismail R, Zaky M. Effect of some natural antioxidants on growth performance, blood parameters and carcass traits of growing rabbits under Egyptian summer condition. Journal of Animal and Poultry Production 2016; 7(12): 457-66.
[http://dx.doi.org/10.21608/jappmu.2016.48755]

[67]   Gonzalez-Rivas PA, Chauhan SS, Ha M, Fegan N, Dunshea FR, Warner RD. Effects of heat stress on animal physiology, metabolism, and meat quality: A review. Meat Sci 2020; 162: 108025.
[http://dx.doi.org/10.1016/j.meatsci.2019.108025] [PMID: 31841730]

[68]   Abdlazez ST, Saleh HH. Effect of diferente doses of propolis on growth performance, some carcass characteristicis and lipid peroxidation stability of Awassi lambs. Al-Anbar Journal of Veterinary Sciences 2016; 9: 62-71.

[69]   Shedeed HA, Farrag B, Elwakeel EA, Abd El-Hamid IS, El-Rayes MAH. Propolis supplementation improved productivity, oxidative status, and immune response of Barki ewes and lambs. Vet World 2019; 12(6): 834-43.
[http://dx.doi.org/10.14202/vetworld.2019.834-843] [PMID: 31440002]

[70]   Çetin E, Silici S, Çetin N, Güçlü BK. Effects of diets containing different concentrations of propolis on hematological and immunological variables in laying hens. Poult Sci 2010; 89(8): 1703-8.
[http://dx.doi.org/10.3382/ps.2009-00546] [PMID: 20634526]

[71]   Shaddel-Tili A, Eshratkhan B, Kouzehgari H, Ghasemi-Sadabadi M. The effect of different levels of propolis in diets on performance, gastrointestinal morphology and some blood parameters in broiler chickens. Bulg J Vet Med 2017; 20(3): 215-24.
[http://dx.doi.org/10.15547/bjvm.986]

[72]   Attia YA, Alagawany MM, Farag MR, *et al.* Phytogenic products and phytochemicals as a candidate strategy to improve tolerance to COVID-19. Front Vet Sci 2020; 7: 573159.
[http://dx.doi.org/10.3389/fvets.2020.573159] [PMID: 33195565]

[73]   Abbas RJ. Effect of dietary supplementation with differing levels of propolis on productivity and blood parameters in broiler chicks. Magallat al-Basrat Li-l-Abhat al-Baytariyyat 2014; 1: 164-80.
[http://dx.doi.org/10.33762/bvetr.2014.98808]

[74]   Silva CSR, Paes Barreto CL, Peixoto RDM, Mota RA, Ribeiro MDF, Da Costa MM. Ação antibacteriana da própolis marrom brasileira em diferentes solventes contra *Staphylococcus* spp. isolados de casos de mastite caprina. Cienc Anim Bras 2012; 13(2): 247-51.
[http://dx.doi.org/10.5216/cab.v13i2.15121]

[75]   Przybyłek I, Karpiński TM. Antibacterial properties of própolis. Molecules 2019; 24(11): 2047.
[http://dx.doi.org/10.3390/molecules24112047] [PMID: 31146392]

[76]   Peter CM, Picoli T, Zani JL, *et al.* Atividade antiviral e virucida de extratos hidroalcoólicos de própolis marrom, verde e de abelhas Jataí (*Tetragonisca angustula*) frente ao herpersvírus bovino tipo 1 (BoHV-1) e ao vírus da diarreia viral bovina (BVDV). Pesqui Vet Bras 2017; 37(7): 667-75.
[http://dx.doi.org/10.1590/s0100-736x2017000700003]

[77]   Yosri N, Abd El-Wahed AA, Ghonaim R, *et al.* Antiviral and immunomodulatory properties of propolis: chemical diversity, pharmacological properties, preclinical and clinical applications, and in silico potential against SARS-CoV-2. Foods 2021; 10(8): 1776.
[http://dx.doi.org/10.3390/foods10081776] [PMID: 34441553]

[78]   Ibrahim MEED, Alqurashi RM. Anti-fungal and antioxidant properties of propolis (bee glue) extracts. Int J Food Microbiol 2022; 361: 109463.
[http://dx.doi.org/10.1016/j.ijfoodmicro.2021.109463] [PMID: 34742143]

[79]   Ożarowski M, Karpiński TM, Alam R, Łochyńska M. Antifungal properties of chemically defined propolis from various geographical regions. Microorganisms 2022; 10(2): 364.
[http://dx.doi.org/10.3390/microorganisms10020364] [PMID: 35208818]

[80]   Prytzyk E, Dantas AP, Salomão K, *et al.* Flavonoids and trypanocidal activity of Bulgarian propolis. J Ethnopharmacol 2003; 88(2-3): 189-93.
[http://dx.doi.org/10.1016/S0378-8741(03)00210-1] [PMID: 12963141]

[81]   Wang BJ, Lien YH, Yu ZR. Supercritical fluid extractive fractionation – study of the antioxidant activities of propolis. Food Chem 2004; 86(2): 237-43.
[http://dx.doi.org/10.1016/j.foodchem.2003.09.031]

[82]   Dias MC, Pinto DCGA, Silva AMS. Plant flavonoids: Chemical characteristics and biological activity. Molecules 2021; 26(17): 5377.
[http://dx.doi.org/10.3390/molecules26175377] [PMID: 34500810]

[83]   Brätter C, Tregel M, Liebenthal C, Volk HD. Prophylactic effectiveness of propolis for immunostimulation: a clinical pilot study. Forsch Komplementarmed 1999; 6(5): 256-60.
[http://dx.doi.org/10.1159/000021260] [PMID: 10575279]

[84]   Dimov V, Ivanovska N, Manolova N, Bankova V, Nikolov N, Popov S. Immunomodulatory action of propolis. Influence on anti-infectious protection and macrophage function. Apidologie (Celle) 1991; 22(2): 155-62.
[http://dx.doi.org/10.1051/apido:19910208]

[85]   Scheller S, Gazda G, Pietsz G, *et al.* The ability of ethanol extract of propolis to stimulate plaque formation in immunized mouse spleen cells. Pharmacol Res Commun 1988; 20(4): 323-8.
[http://dx.doi.org/10.1016/S0031-6989(88)80068-7] [PMID: 3387460]

[86]   Ivanovska N, Neychev H, Stefanova Z, Bankova V, Popov S. Influence of cinnamic acid on lymphocyte proliferation, cytokine release and Klebsiella infection in mice. Apidologie (Celle) 1995; 26(2): 73-81.
[http://dx.doi.org/10.1051/apido:19950201]

[87]   Oršolić N, Bašić I. Antitumor, hematostimulative and radioprotective action of water-soluble derivative of propolis (WSDP). Biomed Pharmacother 2005; 59(10): 561-70.
[http://dx.doi.org/10.1016/j.biopha.2005.03.013] [PMID: 16202559]

[88]   Park JH, Lee JK, Kim HS, *et al.* Immunomodulatory effect of caffeic acid phenethyl ester in Balb/c mice. Int Immunopharmacol 2004; 4(3): 429-36.

[http://dx.doi.org/10.1016/j.intimp.2004.01.013] [PMID: 15037220]

[89]   Kimoto T, Arai S, Kohguchi M, *et al.* Apoptosis and suppression of tumor growth by artepillin C extracted from Brazilian propolis. Cancer Detect Prev 1998; 22(6): 506-15.
[http://dx.doi.org/10.1046/j.1525-1500.1998.00020.x] [PMID: 9824373]

[90]   Borrelli F, Maffia P, Pinto L, *et al.* Phytochemical compounds involved in the anti-inflammatory effect of propolis extract. Fitoterapia 2002; 73 (Suppl. 1): S53-63.
[http://dx.doi.org/10.1016/S0367-326X(02)00191-0] [PMID: 12495710]

[91]   Taheri HR, Rahmani HR, Pourreza J. Humoral immunity of broilers is affected by oil extracted propolis (OEP) in the diet. Int J Poult Sci 2005; 4: 414-7.
[http://dx.doi.org/10.3923/ijps.2005.414.417]

[92]   Kumazawa S, Hamasaka T, Nakayama T. Antioxidant activity of propolis of various geographic origins. Food Chem 2004; 84(3): 329-39.
[http://dx.doi.org/10.1016/S0308-8146(03)00216-4]

[93]   Nagai T, Inoue R, Inoue H, Suzuki N. Preparation and antioxidant properties of water extract of propolis. Food Chem 2003; 80(1): 29-33.
[http://dx.doi.org/10.1016/S0308-8146(02)00231-5]

[94]   Russo A, Longo R, Vanella A. Antioxidant activity of propolis: role of caffeic acid phenethyl ester and galangin. Fitoterapia 2002; 73 (Suppl. 1): S21-9.
[http://dx.doi.org/10.1016/S0367-326X(02)00187-9] [PMID: 12495706]

[95]   Namgoong SY, Son KH, Chang HW, Kang SS, Kim HP. Effects of naturally occurring flavonoids on mitogen-induced lymphocyte proliferation and mixed lymphocyte culture. Life Sci 1994; 54(5): 313-20.
[http://dx.doi.org/10.1016/0024-3205(94)00787-X] [PMID: 8289592]

[96]   Kiaei SM, Modirsanei M, Bozorgmehri H, Mansoori B, Gholamian B, Ghalianehi A. Comparison of the effects of propolis and Virginia mycine on performance and immune response ofbroiler chicks. Majallah-i Tahqiqat-i Dampizishki-i Iran 2008; 62: 367-72.

[97]   Ziaran HR, Rahmani HR, Pourreza J. Effect of dietary oil extract of propolis on immune response and broiler performance. Pak J Biol Sci 2005; 8: 1485-90.
[http://dx.doi.org/10.3923/pjbs.2005.1485.1490]

[98]   Haščík P, Trembecká L, Bobko M, *et al.* Effect of diet supplemented with propolis extract and probiotic additives on performance, carcass characteristics and meat composition of broiler chickens. Slovak Journal of Food Sciences 2016; 10: 223-31. [https://doi.org/10.5219/581].

[99]   Eyng C, Murakami AE, Pedroso RB, Silveira TGV, Lourenço DAL, Garcia AFQM. Crude propolis as an immunostimulating agent in broiler feed during the first stage. Semin Cienc Agrar 2013; 34(5): 2511-22.
[http://dx.doi.org/10.5433/1679-0359.2013v34n5p2511]

[100]   Babaei S, Rahimi S, Karimi Torshizi MA, Tahmasebi G, Khaleghi Miran SN. Effects of propolis, royal jelly, honey and bee pollen on growth performance and immune system of Japanese quails. Vet Res Forum 2016; 7(1): 13-20.
[PMID: 27226882]

[101]   Zafarnejad K, Nazar A, Mostafa R. Effect of bee glue on growth performance and immune response of broiler chickens. Journal of Applied Animal Research 2017; 45: 280-284.
[http://dx.doi.org/10.1080/09712119.2016.1174130]

CHAPTER 12

General Conclusion and Recommendations

Youssef A. Attia[1,2], Mohamed E. Abd El-Hack[3,*], Mahmoud M. Alagawany[3], Mohamed A. AlBanoby[4] and Rashed A. Alhotan[5]

[1] *Sustainable Agriculture Production Research Group, Agriculture Department, Faculty of Environmental Sciences, King Abdulaziz University, Jeddah-21589, Saudi Arabia*

[2] *Animal and Poultry Production Department, Faculty of Agriculture, Damanhour University, Damanhour-22713, Egypt*

[3] *Poultry Department, Faculty of Agriculture, Zagazig University, Zagazig-44519, Egypt*

[4] *Al-Shamel Animal Feed Factory, Industrial Area, Hail-55411, Saudi Arabia*

[5] *Department of Animal Production, College of Food and Agricultural Sciences, King Saud University, Riyadh, Saudi Arabia*

Abstract: The use of phytogenic additives in livestock nutrition as an alternative to classical feed additives has shown promising results in improving animal performance and product quality while avoiding the negative effects of antibiotics on animal health, product quality, and human health. This book reviews the recent advances in photogenic nutrition and its application in animal nutrition as a means of antibiotic replacement and eco-friendly feed additives. This chapter summarizes the outcomes of the 11 chapters reviewed, and their possible applications in animal nutrition. Numerous possible alternatives to antibiotic growth promoters can be used in mono-gastric animal nutrition, including thyme, rosemary, milk thistle seeds, turmeric, phytogenic, essential oils, bee pollen, and propolis. These alternatives and eco-friendly feed additives serve as sources of bioactive ingredients such as flavonoids, phenols, and polyphenols [1–4]. To date, the results have been inconclusive because of the different factors involved in animal responses, such as strain and age of the animal, health conditions, housing conditions, environmental status, part of the plant, type of plant product (leaves, seeds, and roots), drying methods, extraction methods, water *vs.* organic solvents, dose of administration, and methods of delivery (feed and water). Thus, further studies are needed to identify the dose, bioactive substances, and application root to develop commercial products on an individual basis and/or mixed agents that need to be tested. These promising additives may partially or completely replace antibiotic growth promoters and overcome the possible problems caused by the withdrawal of antibiotics from the feed additives market [2, 5, 6]. The use of phytogenic feed supplements in farm animal nutrition as a substitute for classical feed additives has shown promising

* **Corresponding author Mohamed E. Abd El-Hack:** Poultry Department, Faculty of Agriculture, Zagazig University, Zagazig-44519, Egypt; E-mail: dr.mohamed.e.abdalhaq@gmail.com

results in enhancing animal performance and product quality, while avoiding the negative effects of antibiotics on human and animal health, product quality, and food security and safety.

Keywords: Bioactive substances, General recommendations, Livestock, Photogenic.

CONCLUSION AND REMARKS

In conclusion, the use of phytogenic additives in livestock nutrition is a promising alternative to traditional feed additives, offering potential benefits for animal performance and product quality without the drawbacks associated with antibiotics. This review highlights various natural sources of bioactive ingredients such as thyme, rosemary, and milk thistle seeds, which can serve as eco-friendly feed additives. However, the inconsistent results observed across studies underscore the need for further research to address the complex interplay between factors influencing animal responses. Future investigations should focus on optimizing dosages, identifying specific bioactive substances, and developing standardized application methods. Additionally, exploring the long-term effects of phytogenic additives on animal health, product quality, and environmental impacts is crucial. As the livestock industry moves towards more sustainable practices, the development of cost-effective production methods for phytogenic additives is essential. By addressing these research gaps, the potential of phytogenic additives to replace antibiotic growth promoters can be fully realized, contributing to improved food security, safety, and overall animal well-being.

IMPLICATIONS

- Phytogenic additives show promise as alternatives to antibiotic growth promoters, addressing concerns regarding antibiotic resistance and residues in animal products.
- These natural additives may enhance animal growth and the quality of animal products, without the negative effects associated with antibiotics.
- By reducing reliance on antibiotics, phytogenic additives could contribute to improved food security and safety for consumers.
- These inconsistent results across studies highlight the importance of developing standardized protocols for using phytogenic additives, including optimal dosages and application methods.
- The use of eco-friendly feed additives aligns with the growing demand for sustainable livestock production.

- Future studies should focus on the long-term effects, cost-effective production methods, and the potential impact on environmental factors, such as greenhouse gas emissions.
- The variety of phytogenic sources and their bioactive ingredients suggest the potential for developing tailored feed additives for different livestock species and production systems.
- The development of effective phytogenic additives could reshape the feed additive market and affect the economics of livestock production.
- Further research is required to explore the efficacy of these additives across different livestock types and geographical regions.
- Advancing this field requires collaboration among animal nutritionists, plant scientists, and environmental researchers to fully understand and optimize the use of phytogenic additives in livestock production.
- The data in Table **1** summarize natural phytogenic sources, their bioactive ingredients, and their importance for human and animal well-being. The dose of use cannot be suggested herein because of huge differences in the literature according to different experimental conditions, birds, age, and feed strategies.

Table 1. Natural phytogenic sources, their bioactive ingredients, and their importance for human and animal well-being (Adapted from [3, 6 - 9]).

Herbs	Bioactive ingredients in herbs	Benefits concerning health
Turmeric powder	Flavonoids compounds	Antimicrobial, antioxidant
Garlic, onion and their leaves	Allicin, Allyl sulfide	Reduce LDL cholesterol, anticarcinogenic properties
Sugar beet, grape pulp	Betaine	Decrease plasma homocysteine which ruptures arterial walls
Alfaalfa, marigold petals, red pepper, spirulina	Carotenoid pigments	Antioxidant, anticarcinogenic
Basil leaves	Eugenic acid, Eugenol	Immunomodulatory properties
Marigold petals, bay	Lutein	Antioxidants, improve vision
Tomato pomace, grape pulp	Lycopene	Decrease LDL cholesterol, antioxidant, anticarcinogenic
Citrus pulp	Nirangenin	Reduce LDL cholesterol
Flaxseed, canola, fish, oils, insects, worms	n-3 PUFA	Decrease LDL, hypertension, angina, atherosclerosis
Seeds, legumes, weeds, yeast, fermented products	Phytosterols	Increase HDL, decrease blood sugar
Fenugreek, spices	Quercitin, Lutein, Citogenin	Induce insulin secretion, antimicrobial, and tonic activity.

(Table 1) cont.....

Herbs	Bioactive ingredients in herbs	Benefits concerning health
Brewery waste, yeast, fermented products	Statin	Reduce LDL cholesterol
Broccoli, cauliflower, cabbage, radish leaves, waste	Sulforaphane	Anticarcinogenic and antioxidant properties
Bran	Tocotrienols	Decrease LDL cholesterol
Milk, eggs, meat products		Taurine

FUTURE SCOPE

Research should focus on exploring the potential of phytogenic additives in different types of livestock worldwide as well as investigating the long-term effects of phytogenic additives on animal health, product quality, the environment, and green gaseous emissions. Further studies are needed to optimize the use of phytogenic additives in feed formulations as individual or mixed components and to develop cost-effective production methods for sustainable animal production. Further research is needed to identify the dose and type of product using dose-response experiments and broken-line regression analysis. Such research is essential in view of the lack of adequate results regarding the effects of different doses of individual and mixtures of different phytogenic additives.

REFERENCES

[1]   Eevuri T, Putturu R. Use of certain herbal preparations in broiler feeds - A review. Vet World 2013; 6: 172-9.
        [http://dx.doi.org/10.5455/vetworld.2013.172-179]

[2]   Abou-Elkhair R, Ahmed HA, Selim S. Effects of black pepper (*piper nigrum*), turmeric powder (*curcuma longa*) and coriander seeds (*coriandrum sativum*) and their combinations as feed additives on growth performance, carcass traits, some blood parameters and humoral immune response of broiler chickens. Asian-Australas J Anim Sci 2014; 27(6): 847-54.
        [http://dx.doi.org/10.5713/ajas.2013.13644] [PMID: 25050023]

[3]   Abdelli N, Solà-Oriol D, Pérez JF. Phytogenic Feed Additives in Poultry: Achievements, Prospective and Challenges. Animals (Basel) 2021; 11(12): 3471.
        [http://dx.doi.org/10.3390/ani11123471] [PMID: 34944248]

[4]   Shehata AA, Attia Y, Khafaga AF, *et al.* Restoring healthy gut microbiome in poultry using alternative feed additives with particular attention to phytogenic substances: Challenges and prospects. German Journal of Veterinary Research 2022; 2(3): 32-42.
        [http://dx.doi.org/10.51585/gjvr.2022.3.0047]

[5]   Sadeghi GH, Karimi A, Padidar Jahromi SH, Azizi T, Daneshmand A. Effects of cinnamon, thyme and turmeric infusions on the performance and immune response in of 1- to 21-day-old male broilers. Rev Bras Cienc Avic 2012; 14(1): 15-20.
        [http://dx.doi.org/10.1590/S1516-635X2012000100003]

[6]   Shehata AA, Yalçın S, Latorre JD, *et al.* Probiotics, Prebiotics, and Phytogenic Substances for

Optimizing Gut Health in Poultry. Microorganisms 2022; 10(2): 395. b
[http://dx.doi.org/10.3390/microorganisms10020395] [PMID: 35208851]

[7] Lee SH, Lillehoj HS, Jang SI, Lillehoj EP, Min W, Bravo DM. Dietary supplementation of young broiler chickens with *Capsicum* and turmeric oleoresins increases resistance to necrotic enteritis. Br J Nutr 2013; 110(5): 840-7.
[http://dx.doi.org/10.1017/S0007114512006083] [PMID: 23566550]

[8] Attia YA. Mahmoud Alagawany, Mayada R Farag, Asmaa F Khafaga, Abdel-Moneim E Abdel-Moneim, Khalid A Asiry, Noura M Mesalam, Manal E Shafi, Mohammed A Al-Harthi, Mohamed E Abd El-Hack. Phytogenic products and phytochemicals as a candidate strategy to improve tolerance to COVID-19. Front Vet Sci 2020.

[9] Upadhaya, Santi Devi and Kim, In Ho. Efficacy of phytogenic feed additive on performance, production and health status of monogastric animals – a review. Annals of Animal Science 2017; 17, 4, 3917, pp. 929-948.
[http://dx.doi.org/10.1515/aoas-2016-0079]

SUBJECT INDEX

www.ingramcontent.com/pod-product-compliance
Lightning Source LLC
Chambersburg PA
CBHW041442210326
41599CB00004B/99